I0064174

ENGINEERING MICROBIAL METABOLISM FOR CHEMICAL SYNTHESIS

Reviews and Perspectives

ENGINEERING MICROBIAL METABOLISM FOR CHEMICAL SYNTHESIS

Reviews and Perspectives

Editor

Yajun Yan

The University of Georgia, USA

World Scientific

W JERSEY · LONDON · SINGAPORE · BEIJING · SHANGHAI · HONG KONG · TAIPEI · CHENNAI · TOKYO

Published by

World Scientific Publishing Europe Ltd.

57 Shelton Street, Covent Garden, London WC2H 9HE

Head office: 5 Toh Tuck Link, Singapore 596224

USA office: 27 Warren Street, Suite 401-402, Hackensack, NJ 07601

Library of Congress Cataloging-in-Publication Data

Names: Yan, Yajun, editor.

Title: Engineering microbial metabolism for chemical synthesis : reviews and perspectives / edited by Yajun Yan (The University of Georgia, USA).

Description: New Jersey : World Scientific, 2017. | Includes bibliographical references.

Identifiers: LCCN 2017030404 | ISBN 9781786344298 (hc : alk. paper)

Subjects: | MESH: Metabolic Engineering | Microbiological Phenomena

Classification: LCC QP171 | NLM QU 300 | DDC 612.3/9--dc23

LC record available at https://lccn.loc.gov/2017030404

British Library Cataloguing-in-Publication Data

A catalogue record for this book is available from the British Library.

Copyright © 2018 by World Scientific Publishing Europe Ltd.

All rights reserved. This book, or parts thereof, may not be reproduced in any form or by any means, electronic or mechanical, including photocopying, recording or any information storage and retrieval system now known or to be invented, without written permission from the Publisher.

For photocopying of material in this volume, please pay a copying fee through the Copyright Clearance Center, Inc., 222 Rosewood Drive, Danvers, MA 01923, USA. In this case permission to photocopy is not required from the publisher.

For any available supplementary material, please visit
http://www.worldscientific.com/worldscibooks/10.1142/Q0126#t=suppl

Desk Editors: Suraj Kumar/Jennifer Brough/Koe Shi Ying

Typeset by Stallion Press
Email: enquiries@stallionpress.com

About the Editor

Yajun Yan, Ph.D., is an Associate Professor of Biochemical Engineering, in the College of Engineering, University of Georgia, Athens. He obtained his Ph.D. from the State University of New York at Buffalo in the Department of Chemical and Biological Engineering in 2008. During this period, his research was focused on metabolic engineering. From 2008 to 2010, he did his postdoctoral research at the University of California, Los Angeles, with further training in metabolic engineering and synthetic biology. His current research focuses on developing enzymatic and microbial approaches for the production of pharmaceutically important compounds as well as fuels and renewable chemicals.

Contents

Chapter 1

Glycolysis and Its Metabolic Engineering Applications

Jian Wang* and Yajun Yan[†,‡]

*College of Engineering, University of Georgia
Athens, GA 30602, USA
†BioChemical Engineering Program, College of Engineering
University of Georgia, Athens, GA 30602, USA
‡ yajunyan@uga.edu

1.1. Introduction

Microorganisms play a pivotal role in the degradation of complex carbon sources in nature and could survive in different conditions due to the genetically coined cellular metabolism. Cellular metabolism of microorganisms is defined as the sum of biochemical processes that involves the conversion of environmental nutrients into simple biosynthetic building blocks and energy in the cells. Within these metabolic processes, catabolism is responsible for breakdown of complex molecules into simple molecules, producing energy, reducing power, and precursor intermediates for other metabolic processes like anabolism. The central catabolic pathways include Embden–Meyerhof–Parnas (EMP) pathway, pentose phosphate (PP) pathway, Entner–Doudoroff (ED) pathway, and the tricarboxylic acid (TCA) cycle. The EMP pathway, also known as glycolysis

1

pathway, is one of the most ancient metabolic pathways and occurs, if at least in part, in nearly all organisms.[1] Glycolysis pathway is an oxygen-independent metabolic pathway that comprises the initial steps required for the metabolism of carbohydrates like glucose. Its function is essential for cell growth because it generates precursor metabolites from central catabolic pathways that are the starting building blocks for macromolecules and other small molecules. Thus, glycolysis pathway is a gatekeeper pathway for carbohydrate utilization and cell survival. This chapter mainly focuses on the fundamental mechanisms and metabolic engineering applications of glycolysis pathway for the production of a plethora of value-added chemicals in the well-developed *Escherichia coli*.

1.2. The Fundamentals of Glycolysis in Biological Systems

1.2.1. *Glycolysis Pathway*

Glycolysis pathway is one of the main components of central metabolism that is regarded as an entry pathway for glucose utilization. The principal functions of glycolysis pathway are oxidation of hexoses like glucose to generate adenosine triphosphate (ATP), reductants like reduced nicotinamide adenine dinucleotide (NADH) and trioses like pyruvate. During this process, six of 12 precursor metabolites are supplied for biosynthesis (Table 1.1). The reactions of glycolysis pathway can be summarized as a reaction as follows:

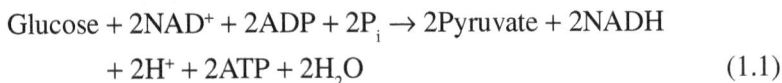

$$\text{Glucose} + 2\text{NAD}^+ + 2\text{ADP} + 2\text{P}_i \rightarrow 2\text{Pyruvate} + 2\text{NADH}$$
$$+ 2\text{H}^+ + 2\text{ATP} + 2\text{H}_2\text{O} \tag{1.1}$$

Glycolysis pathway is a 10-step pathway and can be divided into two stages: the preparatory phase glucose to glyceraldehyde-3-phosphate (G-3-P) and the payoff phase (G-3-P to pyruvate) (Figure 1.1). The preparatory phase consists of the first five steps that include glucose phosphorylation, isomerization, group transfer, and cleavage reactions. Two moles of ATP are utilized and two moles of G-3-P are produced per mole of glucose. The payoff phase consists of subsequent five redox reactions that oxidize G-3-P to pyruvate, during which four moles of ATP and two moles of NADH are produced per mole of glucose. Glycolysis

Table 1.1. Twelve major precursor metabolites generated in central pathways.

Precursor metabolite	Metabolic pathway
Glucose-6-phosphate (G-6-P)	Glycolysis
Fructose-6-phosphate (F-6-P)	Glycolysis
Triose phosphate	Glycolysis
3-Phosphoglycerate (3-PG)	Glycolysis
Phosphoenolpyruvate (PEP)	Glycolysis
Pyruvate	Glycolysis
Ribose-5-phosphate (R-5-P)	PP pathway
Erythrose-4-phosphate (E-4-P)	PP pathway
Acetyl-CoA	TCA cycle
α-Ketoglutarate (α-KG)	TCA cycle
Oxaloacetate	TCA cycle
Succinyl-CoA	TCA cycle

pathway-mediated glucose oxidation is incomplete and the final products (two pyruvates) still contain the bulk of the total energy initially in glucose. Complete oxidation of one mole of glucose will release 2,840 kJ of energy, whereas 94.8% of this energy is stored in two moles of pyruvate, which will be largely released during TCA cycle.[2]

In *E. coli*, the assimilation and phosphorylation of glucose can be realized by the phosphotransferase system (PTS) or by glucokinase (encoded by *glk*). The PTS system, coupled with consumption of one mole of phosphoenolpyruvate (PEP) to pyruvate for each mole of internalized glucose, plays a major role in glucose transport and phosphorylation.[3] The PTS system consists of sugar-non-specific protein components (Enzyme I, encoded by *ptsI*; Hpr, encoded by *ptsH*) and glucose-specific PTS permeases (Enzyme IIA[glc], encoded by *crr*; Enzyme IIBC[glc], encoded by *ptsG*). The phosphate group from PEP is transferred sequentially through Enzyme I, Hpr, Enzyme IIA[glc], Enzyme IIC[glc], and IIB[glc] to glucose.[3] Glucokinase can also produce phosphorylate glucose, but it is not indispensable for cell growth on glucose as the carbon source.[4]

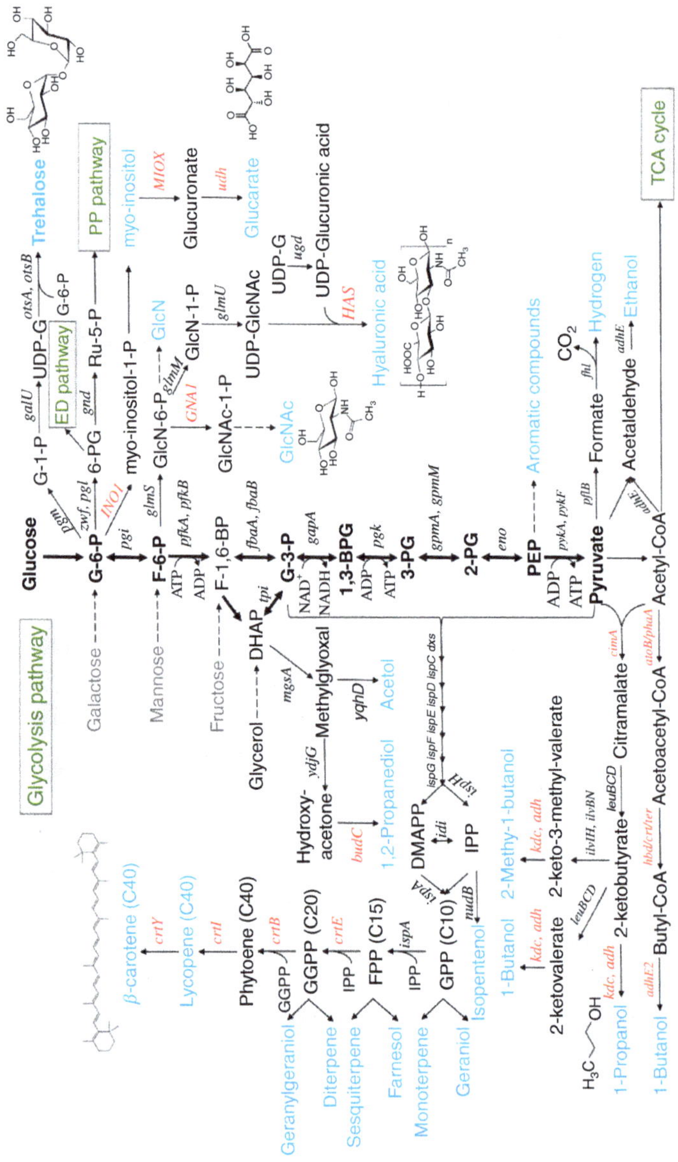

Glycolysis pathway
Glucose
G-6-P
F-6-P
F-1,6-BP
G-3-P
1,3-BPG
3-PG
2-PG
PEP
Pyruvate
Acetyl-CoA
DHAP
Galactose
Mannose
Fructose
Glycerol
Methylglyoxal
Acetol
Hydroxy-acetone
1,2-Propanediol
ED pathway
PP pathway
TCA cycle
Trehalose
G-1-P
UDP-G
6-PG
Ru-5-P
myo-inositol-1-P
myo-inositol
Glucuronate
Glucarate
GlcN-6-P
GlcN
GlcN-1-P
GlcNAc-1-P
GlcNAc
UDP-GlcNAc
UDP-Glucuronic acid
Hyaluronic acid
Aromatic compounds
Formate
Hydrogen
Acetaldehyde
Ethanol
Citramalate
Acetoacetyl-CoA
Butyl-CoA
1-Butanol
2-ketobutyrate
1-Propanol
2-keto-3-methyl-valerate
2-Methy-1-butanol
2-ketovalerate
DMAPP
IPP
GPP (C10)
FPP (C15)
GGPP (C20)
Phytoene (C40)
Lycopene (C40)
β-carotene (C40)
Isopentenol
Geraniol
Monoterpene
Farnesol
Sesquiterpene
Diterpene
Geranylgeraniol

Figure 1.1. Glycolysis pathway and metabolic pathways for various value-added chemicals. Arrows indicate the metabolite flux of the enzymatic reactions and the thicker arrows indicate the glycolysis pathway. Endogenous genes are shown in black and heterologous genes are shown in red. Metabolites referred: G-6-P, glucose-6-phosphate; F-6-P, fructose-6-phosphate; F-1,6-BP, fructose 1,6-bisphosphate; DHAP, dihydroxyacetone phosphate; G3P, glyceraldehyde 3-phosphate; 1,3-BPG, 1,3-bisphosphoglycerate; 3-PG, 3-phosphoglycerate; 2-PG, 2-phosphoglycerate; PEP, phosphoenolpyruvate; G-1-P, glucose-1-phosphate; UDP-glucose, UDP-D-glucose; 6-PG, D-gluconate 6-phosphate; Ru-5-P, ribulose-5-phosphate. GlcN, glucosamine; GlcNAc, N-acetylglucosamine; DMAPP, dimethylallyl diphosphate; IPP, isopentenyl diphosphate; GPP, geranylgeranyl phosphate; FPP, farnesyl pyrophosphate; GGPP, geranylgeranyl diphosphate. Enzymes referred: *pgi*, phosphoglucose isomerase; *pfk*, 6-phosphofructokinase; *fba*, fructose bisphosphate aldolase; *tpi*, triose phosphate isomerase; *gapA*, glyceraldehyde-3-phosphate dehydrogenase A; *pgk*, phosphoglycerate kinase; *gpm*, phosphoglycerate mutase; *eno*, enolase; *pfk*, pyruvate kinase; *zwf*, glucose-6-phosphate-1-dehydrogenase; *pgl*, 6-phosphogluconolactonase; *gnd*, 6-phosphogluconate dehydrogenase; *INO1*, myo-inositol-1-phosphate synthase; *MIOX*, myo-inositol oxygenase; *udh*, uronate dehydrogenase; *pgm*, phosphoglucomutase; *galU*, UTP:glucose-1-phosphate uridylyltransferase; *otsA*, trehalose-6-phosphate synthase; *otsB*, trehalose-6-phosphate phosphatase; *glmS*, GlcN-6-P synthase; *glmM*, phosphoglucosamine mutase; *glmU*, fused N-acetylglucosamine-1-phosphate uridyltransferase and glucosamine-1-phosphate acetyltransferase; *GNA1*, GlcN-6-P N-Acetyltransferase; *ugd*, UDP-glucose 6-dehydrogenase; *HAS*, hyaluronic acid synthase; *mgsA*, methylglyoxal synthase; *yqhD*, aldehyde oxidoreductase; *ydjG*, methylglyoxal reductase; *budC*, secondary alcohol dehydrogenase; *pfl*, pyruvate formate-lyase; *fhl*, formate hydrogenlyase; *adhE*, aldehyde/alcohol dehydrogenase; *cimA*, citramalate synthase; *leuA*, 2-isopropylmalate synthase; *leuB*, 3-isopropylmalate dehydrogenase; *leuCD*, isopropylmalate isomerase; *ilvIH*, acetohydroxybutanoate synthase/acetolactate synthase; *ilvBN*, acetohydroxy acid synthase; *kdc*, keto acid decarboxylase; *adh*, alcohol dehydrogenase; *phaA*, acetoacetyl-CoA thiolase; *atoB*, thiolase; *ter*, trans-enoyl-CoA reductase; *crt*, crotonese; *hbd*, 3-hydroxybutyryl-CoA dehydrogenase; *nudB*, phosphatase; *dxs*, 1-deoxy-D-xylulose 5-phosphate synthase; *dxr*, 1-deoxy-D-xylulose 5-phosphate reductoisomerase; *ispF*, 2C-methyl-D-erythritol 2,4-cyclodiphosphate synthase; *ispG*, 1-hydroxy-2-methyl-2-(E)-butenyl 4-diphosphate synthase; *ispH*, 1-hydroxy-2-methyl-2-(E)-butenyl 4-diphosphate reductase; *idi*, isopentenyl diphosphate isomerase; *ispA*, FPP synthase; *crtE*, GGPP synthase; *crtB*, phytoene synthase; *crtI*, phytoene desaturase; *crtY*, lycopene cyclase.

Phosphoglucose isomerase (encoded by *pgi*) catalyzes the isomerization of glucose-6-phosphate (G-6-P) to fructose-6-phosphate (F-6-P), an essential step of the glycolysis pathway. Two 6-phosphofructokinase isozymes, Pfk I and Pfk II (encoded by *pfkA* and *pfkB*), do not share sequence similarity and catalyze the phosphorylation of F-6-P on the C1 carbon with the involvement of ATP consumption. More than 90% of the phosphofructokinase activity are attributed to Pfk I and only less than 5% are attributed to Pfk II.[5] *E. coli* also contains two classes of fructose-1,6-bisphosphate aldolases, FbaA and FbaB (encoded by *fbaA* and *fbaB*), which catalyze a reversible aldol cleavage/condensation reaction of fructose 1,6-bisphosphate (FBP) during glycolysis and gluconeogenesis.[6] The class II aldolase FbaA (metallo) utilizes a divalent metal ion zinc while the class I aldolase FbaB (Schiff base) utilizes a catalytic lysine residue (Lys236) to stabilize the catalytic intermediate. FbaA is required for glycolysis while FbaB most likely participates in gluconeogenesis as it is not expressed when grown on glucose but induced when grown on C3 carbon sources.[7,8] In the glycolytic direction, FBP is cleaved to produce dihydroxyacetone phosphate (DHAP) and G-3-P. Triosephosphate isomerase (TpiA) (encoded by *tpi*) catalyzes isomerization between G-3-P and DHAP. G-3-P dehydrogenase A (encoded by *gapA*) catalyzes the reversible oxidative phosphorylation of G-3-P to 1,3-bisphosphoglycerate (1,3-BPG) in the presence of NAD+. In the glycolytic direction, two moles of NADH will be produced per mole of glucose consumed. Phosphoglycerate kinase (Pgk; encoded by gene *pgk*) is responsible for the reversible phosphorylation between 3-phosphoglycerate (3-PG) and 1,3-BPG. In the glycolytic direction, Pgk catalyzes the transfer of a phosphoryl group from 1,3-BPG to ADP, forming ATP and 3-PG. *E. coli* contains both a 2,3-bisphosphoglyerate-dependent phosphoglycerate mutase GpmA (encoded by *gpmA*) and a cofactor-independent phosphoglycerate mutase GpmM (encoded by *gpmM*), which catalyze the interconversion between 3-PG and 2-phosphoglycerate (2-PG). The GpmA enzyme has significantly higher specific activity than GpmM.[9,10] Enolase (encoded by eno) catalyzes the interconversion of 2-PG and PEP. Two pyruvate kinases, PykA and PykF (encoded by *pykA* and *pykF*), are key and last enzymes of the glycolysis pathway, catalyzing the irreversible transfer of the phosphoryl group of PEP to ADP to form pyruvate and generate ATP.

Pyruvate then will be either used in different metabolic pathways, or go into TCA cycle for further oxidization.

Glycolysis pathway is an amphibolic pathway because it can reversibly produce hexoses from various low-molecular-weight molecules, which is also known as gluconeogenesis. The bidirectionality is primarily owing to the reversibility of most of the enzymatic reactions comprising glycolysis pathway. The reactions catalyzed by the two exceptions [6-phosphofructokinases (PFK) and pyruvate kinases (PyK)] are rendered functionally reversible by other enzymes [fructose-1,6-bisphosphatases Fbp, GlpX, YggF, and YbhA, and phosphoenolpyruvate synthetase (PpsA)]. Thus, various substrates fed into glycolysis at different branch points might lead forward to glycolysis or reverse to gluconeogenesis. This depends on the carbon source being utilized and in order to meet the cell's demand of precursor metabolites for catabolism or biosynthesis. Hexoses like fructose, galactose, mannose, or triose like glycerol can serve as the sole source of carbon for *E. coli*, because they can be internalized by *E. coli* using different mechanisms and feed into glycolysis at different branch points (Figure 1.1). For example, fructose is taken up via the fructose PTS permease, entering the cell in the form of fructose-1-phosphate (F-1-P). F-1-P is then phosphorylated by 1-phosphofructokinase (encoded by *fruK*) at the C-6 carbon to generate fructose-1,6-bisphosphate (F1,6-BP), which can enter glycolysis pathway. Glycerol assimilation in *E. coli* is mediated by the membrane glycerol diffusion facilitator. Internalized glycerol is phosphorylated by glycerol kinase (encoded by *glpK*) to glycerol-3-phosphate (glycerol-3-P) with ATP as the phosphoryl donor. Cytoplasmic glycerol-3-P can be imported into the cell by the GlpT transporter and can be further converted to DHAP by either of two membrane-bound glycerol-3-P dehydrogenase (GlpABC or GlpD), depending on the aerobic or anaerobic conditions. The DHAP can flow into either the direction of glycolysis for energy conversion or into the reverse direction of gluconeogenesis for biosynthesis like peptidoglycan formation.

Glycolysis pathway is connected with other central metabolic pathways at certain branching nodes. The PP and ED pathway branch respectively from gluconate-6-phosphate (6-PG) and ribulose-5-phosphate (Ru-5-P), which are both extended from G-6-P. PP pathway will direct carbon flux back to G-3-P while ED pathway will direct carbon flux back to G-3-P and pyruvate. TCA cycle is a downstream pathway of glycolysis pathway, which is initiated from decarboxylation of pyruvate to acetyl-CoA. ED pathway is recognized as the most common bacterial alternative glycolytic pathway, because its general scheme is quite similar with EMP pathway that involves phosphorylation of glucose and cleavage into two three-carbon units to release ATP.[11] However, EMP and ED pathways differ slightly in the specific redox cofactors they use (e.g., NAD^+ vs. $NADP^+$) and prominently in ATP yield. The EMP pathway produces two moles of ATP per mole of glucose while the ED produces only one mole.[12] Interconnection of multiple metabolic pathways enables bacteria to respond quickly to fluctuations of intracellular metabolites or changes of exogenous growth conditions.

1.2.2. *Regulation of Glycolysis*

Enzymes in the glycolysis pathway are subject to complex control, in order to realize the switch between glycolysis and reverse gluconeogenesis in response to regulators or metabolites. These regulation mechanisms include (1) transcription activation or repression via transcriptional regulators, (2) post-transcriptional regulation via control of mRNA stability, and (3) allosteric regulation mediated by metabolites (Figure 1.2). A set of global transcriptional regulators are involved in glycolysis, such as cAMP receptor protein (Crp), catabolite repressor activator (Cra), and superoxide response protein (SoxR) and Mlc protein. Crp could activate transcription of glycolytic genes like *ptsHI*, *fbaA*, *gapA*, *pgk*, SoxS activates *pgi*, and Cra activates *pps*. But in most cases, Cra is functioning as a repressor of glycolytic genes like *pgk*, *gpmA*, *gpmM*, *eno*, *pykA*, and *pykF*. Another repressor Mlc was revealed to repress sugar transport genes including *ptsG, crr,* and *ptsHI*.[13] Carbon storage regulator (CsrA) is a RNA-binding protein that can prevent the translation of target mRNA via sequestering and/or facilitating mRNA degradation and can also realize gene activation via stabilizing target transcript.[14] CsrA system is a global regulator of a lot of targets in the glycolysis pathway with a positive effect for *pgi*, *pfkA*, *tpiA*, *eno*, *pykA*, and *pykF* and a negative effect for *pgm*, *pfkB*, and *pps*. Thus, attenuation

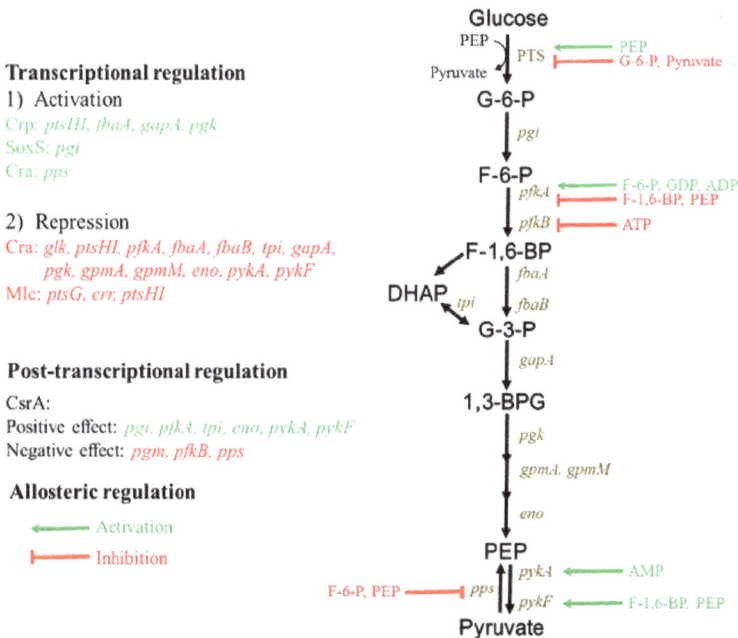

Figure 1.2. Mechanisms of regulation involved in glycolysis pathway. Transcriptional regulation, post-transcriptional regulation, and allosteric regulation are shown as indicated. Green arrows denote activation and red crosses denote inhibition by intracellular metabolites.

of CsrA activity results in a decrease in most glycolytic activities, especially the phosphofructokinase. However, repression of CsrA via synthetic sRNA was found to increase tyrosine production, which was initiated from PEP and E-4-P.[15] Many enzymes in glycolysis pathway are subject to allosteric control, often with inhibitors and activators binding to the same effector site. For example, phosphofructokinase I (Pfk I) is inhibited by PEP and F-1,6-BP but activated by F-6-P, ADP and GDP. Mutation of a single residue in the effector site (Glu187 to Ala) leads to PEP being an activator rather than an inhibitor.[16] Other allosteric enzymes include PTS system (activated by PEP while inhibited by G-6-P and pyruvate), Pfk II (inhibited by ATP), PykA (activated by AMP), PykB (activated by F-1,6-BP and PEP), and Pps (inhibited by F-1,6-BP and PEP).

1.3. Metabolic Engineering Applications of Glycolysis for Microbial Chemical Synthesis

Glycolysis provides a pathway for assimilation and degradation of glucose, during which process multiple metabolites could be utilized as precursors for production of value-added products. These products include saccharides, glucaric acid, glucosamine, aromatic compounds, terpenoids, and biofuels like alcohol and hydrogen.

1.3.1. *Saccharides*

G-6-P and F-6-P, produced in the preparatory phase of glycolysis pathway, can be used for the biosynthesis of commercially valuable oligosaccharides or polysaccharides in *E. coli*. Trehalose is a non-reducing disaccharide with a huge market as an alternative sweetener, water retainer in cosmetics, preservative in pharmaceutical products and frozen foods, and organ protectants for transplants.[17,18] The pathway of trehalose biosynthesis starts at a branch point in G-6-P where fluxes are directed to UDP-glucose via PGM (phosphoglucomutase) and GalU (UTP: glucose-1-phosphate uridylyltransferase) (Figure 1.1). Then, G-6-P and UDP-glucose are coupled to produce trehalose via the function of trehalose-6-phosphate synthase (OtsA) and trehalose-6-phosphate phosphatase (OtsB). Hence, flux distributions between G-6-P and UDP-glucose will impact end-product accumulation. However, overexpression of *otsBA* resulted in low production of trehalose because of concomitant trehalose degradation by endogenous trehalases (TreA and TreF) and trehalose-6-phosphate hydrolase (TreC) in *E. coli*.[19,20] Overexpression of *otsBA* along with gene disruption of *treA*, *treF*, and *treC* or inhibition of trehalases in the presence of validamycin A could effectively increase production and accumulation of trehalose.[19,21] Specially, a five-step pathway for trehalose synthesis was established and optimized on a functional protein chip that achieved *in vitro* production of trehalose.[18]

2'-O-fucosyllactose (2'-FL), a synthetic trisaccharide and one of the most abundant oligosaccharides in human milk, can be potentially used as nutritional additives in infant formula and food supplements for special medical purposes for children and adults. 2'-FL can be produced via overexpression of the fucosyltransferase (FucT2) from *Helicobacter pylori* in GDP-L-fucose producing *E. coli*.[22,23] With additional deletion of endogenous lactose operon and adding three aspartate molecules at the N-terminal of FucT2, 6.4 g/L of 2′-FL was produced in engineered *E. coli* BL21 star (DE3).[24]

Hyaluronic acid (HA) is a glucuronic acid-derived polysaccharide discovered in many animal tissues, which has remarkable applications in pharmaceutical, biomedical, and cosmetic products with high commercial value.[25] Large-scale production of HA has focused on direct extraction from animal tissues and the use of bacterial expression systems in *Streptococci*.[25] *E. coli* has also been developed as a microbial factory for HA production via expression of hyaluronic acid synthase (HAS) gene from *Streptococcus pyogenes*, UDP-glucose 6-dehygrogenase (Ugd), Glucose-1-P uridyltransferase (GalF), and *N*-acetyl glucosamine uridyl-transferase (GlmU) from *E. coli* (Figure 1.1).[26] Heparin and heparan sulfate (HS), chondroitin, and chondroitin sulfate are important glycosamino-glycan biopolymers commonly used in pharmaceutical applications owing to their anticoagulation, antiinflammatory, anticancer, antimeta-static, and antiangiogenic properties.[27–29] Instead of extraction from animal tissues, animal polysaccharides like chondroitin and heparosan have also been engineered in microbial hosts.[30] The biosynthetic pathway of heparosan, the precursor of heparin, has been engineered in non-pathogenic *E. coli* BL21 via expression of *kfiABCD* operon and heparosan was produced with a titer of 1.88 g/L in 3-L bioreactor.[31] The chondroitin pathway consisting of three genes *kfoA* (encoding uridine diphosphate (UDP)-GlcNAc 4-epimerase), *kfoC* (chondroitin polymerase), and *kfoF* (UDP-glucose dehydrogenase) from *E. coli* K4 was introduced into the non-pathogenic *E. coli* BL21Star™ (DE3) and 2.4 g/L chondroitin was produced in 2 L bioreactor.[32]

1.3.2. Glucaric Acid

Glucaric acid is widely recognized as a "top value-added chemical from biomass" and can be used as a building block for polymers for therapeutic purposes like cholesterol reduction, and antidiabetes and anticancer appli-cations.[33–35] The intermediate of glucaric acid, *myo*-inositol, can be used as a precursor for synthesis of *scyllo*-inositol, which has been regarded as a potential therapeutic for Alzheimers.[36,37] Production of glucaric acid is a three-step pathway initiated from G-6-P from glycolysis pathway (Figure 1.1). Establishment of glucaric acid biosynthetic pathway via coexpression of *myo*-inositol-1-phosphate synthase (INO1) gene from *Saccharomyces cerevisiae*, *myo*-inositol oxygenase (MIOX) gene from mice, and urinate dehydrogenase (Udh) gene from *Pseudomonas syringae* in *E. coli* enabled production of glucaric acid via the intermediates *myo*-inositol and glucuronic acid.[38] MIOX was identified as rate-limiting enzyme in the pathway because of accumulation of *myo*-inositol and glucuronic acid and the MIOX activity was strongly influenced by the concentration of the *myo*-inositol.[38,39] Specifically, synthetic protein scaffolds were created via colocalizing all three heterologous enzymes (INO1, MIOX, and Udh) in a designable manner to improve the effective concentration of *myo*-inositol.[39] The synthetic scaffolds significantly increased the specific activity of MIOX and resulted in a 5-fold improvement of glucaric acid production titer (2.5 g/L) over the non-scaffolded control.[39] Another strategy improved the MIOX solubility via an N-terminal SUMO fusion tag and increased *myo*-inositol transport via an insertion of manX fragment in expression vector, which in combination improved production of up to 4.85 g/L of glucaric acid from 10.8 g/L *myo*-inositol.[35] In a recent study, dynamic knockdown via controllable degradation of a key glycolytic enzyme phosphofructokinase-I (Pfk-I) led to an increased G-6-P pool and a 2-fold improvement in yield and titers of *myo*-inositol.[40]

1.3.3. *Glucosamine*

Glucosamine (GlcN, 2-amino-2-deoxy-D-glucose) and *N*-acetylglucosamine (GlcNAc, 2-acetamido-2-deoxy-D-glucose) are amino sugars that have been utilized in dietary supplement and pharmaceutical industries with a market estimated to be at $2 billion.[41] Especially, GlcN has been used in clinical trials to treat osteoarthritis.[42–44] GlcN and GlcNAc are also precursors for glycosaminoglycans like HA and chondroitin sulfate. Acid or enzymatic hydrolysis of chitin is a major route to obtain GlcN and GlcNAc.[45] A microbial pathway for GlcNAc production consists of two enzymes, an *E. coli* glucosamine synthase (GlmS) that catalyzes the synthesis of glucosamine-6-phosphate and a *Saccharomyces cerevisiae* glucosamine-6-phosphate acetyltransferase (GNA1) that converts glucosamine-6-phosphate into *N*-acetylglucosamine.[46] *E. coli* recombinant strain was manipulated with the deletion of GlcN and GlcNAc degradative pathways via mutation of *manXYZ* operon and deleting the *nag* regulon.[44] Expression of GlmS variants resistant to product inhibition by glucosamine-6-P in engineered *E. coli* host led to a GlcN production of 17 g/L and GlcNAc production of 110 g/L in 1 L fermenters.[44,47] Besides strain optimization and enzyme engineering, the culture conditions and dissolved oxygen (DO) levels are also important for the high-level production of GlcN and GlcNAc. A multistage glucose supply strategy or a stepwise DO control strategy enhanced the total GlcN and GlcNAc yields to 69.66 and 72.89 g/L, respectively.[42,48]

1.3.4. *Aromatic Compounds*

Aromatic amino acids, including phenylalanine (Phe), tyrosine (Tyr), and tryptophan (Trp), are essential amino acids for human diets and are important precursors for synthesis of a plethora of high-value chemicals like flavor ingredient (e.g., 2-phenylethanol and vanillin), antioxidants (e.g., coumaric acid, flavonoids, caffeic acid, and its derived esters and amides), and antidepressants (e.g., serotonin and 5-hydroxytryptophan).[49-56] Aromatic compounds are produced via shikimate pathway in bacteria and their biosynthesis is initiated by the condensation of erythrose 4-phosphate (E4P) and PEP to form 3-deoxy-D-arabino-heptulosonate-7-phosphate (DAHP), which will be further converted to chorismate and subsequently branched into Phe, Tyr, and Trp. Thus, metabolic engineering of aromatic amino acids has first focused on enhancing carbon flux to chorismate. Overexpression of transketolases especially (TktA, encoded by *tktA*) is most efficient in increasing the availability of E4P. The availability of PEP is limited due to many competing pathways consuming PEP, including sugar transport by the PTS system (50%), glycolysis (15%), peptidoglycan synthesis (16%), anapleurotic pathway (16%), and shikimate pathway (3%).[57] Deletion of PTS system will theoretically significantly increase the supply of PEP. However, inactivation of the *ptsHI-crr* operon (PTS–strains) would impair cell growth because of severely reduced capacity of glucose transport. Selection of *E. coli* strains, with inactivated PTS, but capable of internalizing glucose with GalP and glucokinase, enabled cells to grow rapidly using glucose as the sole carbon source and increase availability of PEP and production of aromatic compounds.[58] Another strategy is expressing native *galP* and *glk* or *glfZ*$_m$ (glucose facilitator) and *glkZ*$_m$ (glucokinase) from *Zymomonas mobilis* in PTS−strains.[57,59] Overexpression of native PEP synthetase *ppsA* will recycle pyruvate back to PEP pool and enhance availability of PEP for aromatic pathway.[60,61] In *E. coli*, carbon storage regulator protein CsrA is global regulator that negatively impacts PEP synthesis by repressing *pckA* and *ppsA* while activating *pykF*, which channels flux away from PEP.[14,62,63] Repression or disruption of *csrA* will result in increased levels of PEP and increased production of aromatic amino acids.[15,61] Deletion of CsrA, additional overexpression of feedback-inhibition-resistant aromatic pathway enzymes like AroG^{fbr} (DAHP synthase) and TyrA^{fbr} (chorismate mutase/prephenate dehydrogenase), and/or deletion of transcriptional repressor gene *tyrR* or *trpR*, will further enhance the production of aromatic compounds.[60]

1.3.5. *Pyruvate*

Pyruvate is widely used in food additives, nutraceuticals, pharmaceuticals, and as precursors for the synthesis of amino acids like alanine, valine, and aromatic amino acids.[64,65] Pyruvate is the end metabolite intermediate in glycolysis pathway and a critical branch point for TCA cycle, lactate, and acetate production (Figure 1.3). To enhance and accumulate pyruvate in *E. coli*, systematic engineering of *E. coli* host was performed.[65] An improved productive strain *E. coli* TC44 was generated by combining mutations to minimize ATP yield, cell growth, and CO_2 production ($\Delta atpFH$, $\Delta sucA$) with mutations that eliminate acetate production ($\Delta poxB$, $\Delta ackA$) and fermentation products ($\Delta focA$, $\Delta pflB$, $\Delta frdBC$, $\Delta ldhA$, $\Delta adhE$) (Figure 1.3). The final strain could convert glucose to pyruvate with a yield of 0.75 g/g glucose.[65] A similar *E. coli* strain ALS1059 with combined mutations of *aceEF*, *pfl*, *poxB*, *pps*, *ldhA*, *atpFH*, and *arcA* produced 90 g/L pyruvate with a yield of yield of 0.68 g/g glucose.[64] To enhance lactate production from pyruvate, an engineered *E. coli*

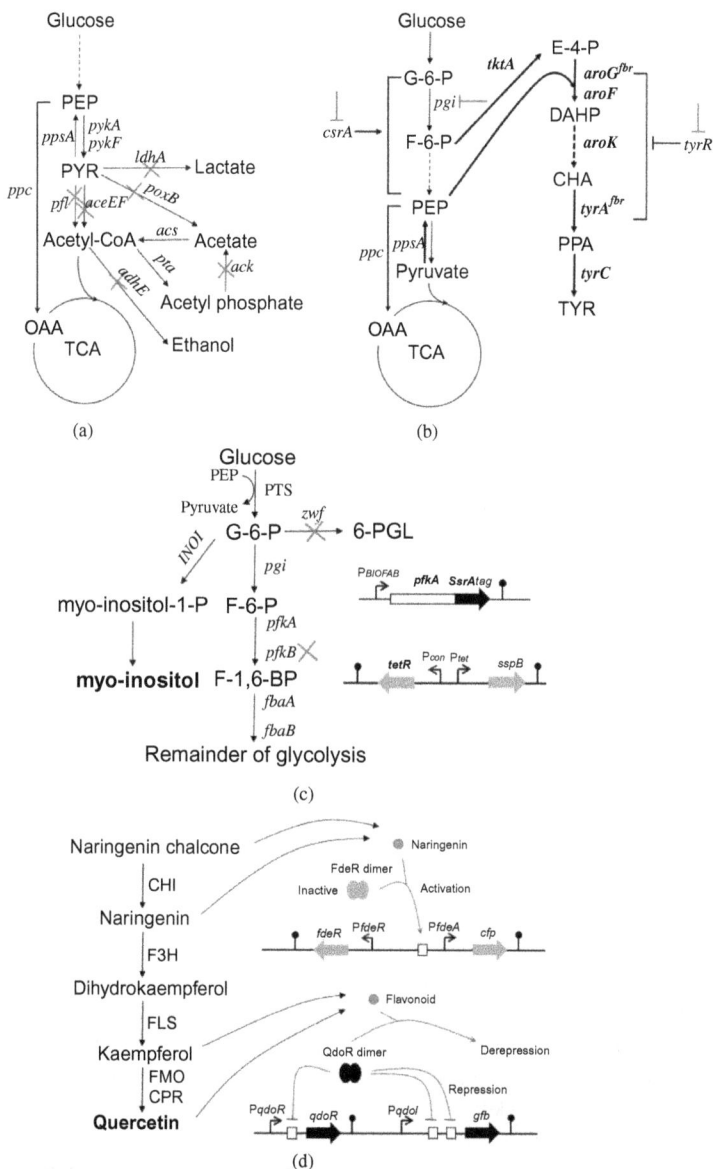

Figure 1.3. Various strategies in metabolic engineering of glycolytic pathway for chemical production in *E. coli*. (a) Pyruvate production via systematic gene deletion. Genes referred: *ldhA*, lactate dehydrogenase; *poxB*, pyruvate oxidase; *pfl*, pyruvate formate lyase; *aceEF*, pyruvate dehydrogenase complex; *acs*, acetyl-CoA synthetase; *ack*, acetate kinase;

strain ALS974 with multigene mutations of *aceEF, pfl, poxB, pps,* and *frdABCD* was developed to achieve 138 g/L lactate production with a yield of 0.99 g/g carbon source.[66]

1.3.6. *Alcohols*

The diminishing fossil fuel reserves and increasing environmental concerns has driven an urgent demand for biofuels from biorenewable resources.[67] In particular, microbial-based bioethanol production has drawn much attention, with more than 3.95 billion liters of ethanol being produced from feedstocks in the United States in 2009.[68] Ethanol biosynthesis is initiated from decarboxylation of pyruvate. Integration of alcohol dehydrogenase II (*adhB*) and pyruvate decarboxylase (*pdc*) from *Zymomonas mobilis* into *E. coli* enabled the production of ethanol from glucose or sugar mixtures.[69,70] Later, a lactate producing *E. coli* was reprogrammed for ethanol production with a titer of 45 g/L by deleting genes encoding all fermentative routes (Δ*focA,* Δ*pflB,* Δ*frd,* Δ*ldhA,* Δ*adhE,* and

---

Figure 1.3. (*Continued*) *pta*, phosphotransacetylase. Deleted genes are indicated in cross. (b) The scheme of engineering of tyrosine biosynthetic pathway via synthetic sRNA. Overexpressed genes are shown in bold and synthetic sRNA repressed targets are shown in vertical cross lines. PYR, pyruvate; E-4-P, erythrose-4-phosphate; DAHP, 3-deoxy-D-arabino-heptulosonate 7-phosphate; CHA, chorismate; PPA, prephenate; TYR, tyrosine. Overexpressed genes are *ppsA*, phosphoenolpyruvate synthase; *tktA*, transketolase A; *aroF*, DAHP synthase; *aroG^fbr*, DAHP synthase with a D146N substitution; *aroK*, shikimate kinase I; *tyrA^fbr*, chorismate mutase/prephenate dehydrogenase with M53I and A354V substitutions; *tyrC*, prephenate dehydrogenase originated from *Zymomonas mobilis*. Synthetic sRNA repressed targets are *tyrR*, tyrosine repressor; *csrA*, carbon-storage regulator; *pgi*, phosphoglucose isomerase. (c) Increase of *myo*-inositol production via dynamic knockdown. Gene knockouts of *zwf* and *pfkB* (Pfk-II) made Pfk-I the sole control point for G-6-P utilization. Further modifications include fusing the *pfkA* with SsrA tag and expressing *sspB* under the control of TetR, generating a Tc-inducible control of Pfk-I degradation. (d) Scheme of the flavonoid synthetic pathway from naringenin chalcone to quercetin and the biosensors to monitor the *in vivo* production of quercetin from naringenin. CHI, chalcone isomerase; F3H, flavanone 3-hydroxylase; FLS, flavonol synthase; FMO, flavonoid 3′-monooxygenase; CPR, NADPH-cytochrome P450 reductase.

Δ*ackA*) for NADH and integrating the complete *Z. mobilis* ethanol pathway (*pdc*, *adhA*, and *adhB*) into the chromosome.[71] Systematic engineering of *E. coli* enabled efficient production of ethanol with a titer of 38.81 g/L from hexoses and pentoses via minimization of the functional space of the central metabolic network.[72]

Higher alcohols (C3 or higher) have been considered as an attractive gasoline substitute because of higher energy densities, less hygroscopicity, and are less volatile than ethanol.[73] Several higher alcohols have been produced via carbon extension and keto acid decarboxylation pathway from pyruvate like 1-propanol, 1-butanol, isobutanol, 2-methyl-1-butanol, and 3-methyl-1-butanol in *E. coli* (Figure 1.1).[73] 1-Propanol was produced via decarboxylation 2-ketobutyrate, which was obtained from coupling of pyruvate and acetyl-CoA via the citramalate synthase (CimA) from *Methanococcus jannaschii*.[74] Further extension of 2-ketobutyrate to 2-ketovalerate or 2-keto-3-methyl-valerate via leucine pathway (LeuBCD) or isoleucine pathway (IlvIHBN) enabled production of 1-butanol and 2-methyl-1-butanol. With CimA variant (CimA 3.7) screened by directed evolution, 3.5 g/L 1-propanol and 0.524 g/L 1-butanol were produced after 92 h.[74] Two novel pathways for 1-propanol biosynthesis were later established. One pathway was designed by expanding the well-known 1,2-propanediol pathway from methylglyoxal shunt with two more enzymatic steps catalyzed by a 1,2-propanediol dehydratase and an alcohol dehydrogenase.[75] The other novel pathway was achieved via extension of threonine pathway from oxaloacetate and decarboxylation of 2-ketobutyrate.[76] Synergy of citramalate pathway and threonine pathway achieved 1-propanol production with a titer of 8 g/L after 72 h.[77] A novel 1-butanol was also established via a six-step enzymatic pathway from acetyl-CoA to butyryl-CoA and 1-butanol (Figure 1.1).[78] Isobutanol was produced via decarboxylation of 2-ketoisovalerate from valine biosynthesis pathway.[73] *E. coli* strain was screened for mutants with the ability to grow with the valine analog norvaline, and one *E. coli* strain NV3r1 was obtained. A final isobutanol titer of 21.2 g/L was achieved with a yield of 0.31 g/g glucose.[79] Modulating the expression of transhydrogenase (*pntAB*) and NAD kinase (*yfjB*) genes for increasing NADPH supply enhanced isobutanol production to a titer of 10.8 g/L with a yield of 0.62 mol/mol.[80] 3-Methyl-1-butanol (3MB) was produced from the decarboxylation of

2-keto-4-methyl-pentanoate from leucine pathway.[73] Random mutagenesis with leucine analogue 4-aza-D,L-leucine resulted in a new *E. coli* strain mutant that was able to produce 4.4 g/L of 3MB.[81]

1.3.7. *Hydrogen*

Biohydrogen (H_2) production by microbes provides an alternative and emerging solution to conventional hydrogen production like electrolysis of water and refining from oil or natural gas, which enables a cheap and feasible process for a clean fuel production from renewable sources.[82,83] Generally, biohydrogen production is carried out mainly through two ways in microorganisms: photosynthesis via photosynthetic microorganisms like *Chlamydomonas reinhardtii* and *Rhodobacter sphaeroides* and anaerobic fermentation via *E. coli*, *Enterobacter*, and *Clostridium* species.[84] *E. coli* can produce H_2 from pyruvate via the successive catalysis of pyruvate formate lyase (PFL) and formate hydrogen lyase (FHL) (Figure 1.1).[85–87] The greatest challenge for fermentative H_2 production is that the yield is low.[88] To maximize H_2 production rate and yield, genetically engineered *E. coli* strains were developed via deletion of *hycA* (a negative regulator for FHL), *hya* and *hyb* (two uptake hydrogenases), *ldhA* (lactate dehydrogenase), and *frdAB* (fumarate reductase).[89] The final engineered strain improved the H_2 yield to 1.80 mol/mol glucose under high H_2 pressure or 2.11 mol/mol glucose under reduced H_2 pressure.[89] Engineering a synthetic NAD(P)H:H_2 pathway in *E. coli* BL21(DE3) achieved H_2 yield to 5.2 mol/mole glucose.[90] Recently, a systematic engineering of *E. coli* host for glycerol fermentation via knockouts of seven genes (*frdC*, *ldhA*, *fdnG*, *ppc*, *narG*, *mgsA*, and *hycA*) created the best strain that reached the theoretical maximum yield of 1 mol H_2/mol glycerol.[91] Overexpression of *zwf* and *gnd* in *E. coli* BW25113 mutant ($\Delta hycA\Delta hyaAB\Delta hybBC\Delta ldhA\Delta frdAB\Delta pfkA$) achieved coproduction of ethanol and H_2 with yields of 1.88 and 1.40 mol/mol glucose, respectively.[92] Although genetic manipulations and utilization of different carbohydrates have been applied to improve hydrogen yield in *E. coli*, the hydrogen production is still in the laboratory scale and additional efforts should be made for scaling up hydrogen production to industrial levels.

1.3.8. *Terpenoids*

Terpenoids, also known as isoprenoids, are plant-derived phytochemicals mostly rendered with antiinflammatory, antiinfectious, and even anticancer properties.[93,94] Complex terpenoids are synthesized from two main C_5 units: isopentenyl pyrophosphate (IPP) and dimethylallyl pyrophosphate (DMAPP). Extension of these C_5 units enabled the production of monoterpenes and geraniol (C_{10}), sesquiterpenes and farnesol (C_{15}), diterpenes and geranylgeraniol (C_{20}), triterpenes (C_{30}), phytoene, lycopene, and carotene (C_{40}) (Figure 1.1). Novel natural or non-natural C_{50} carotenoids like C_{50}-sarcinaxanthin and C_{50}-astaxanthin have also been produced in *E. coli* hosts.[95,96] Terpenoid precursors are generated via two pathways: the 1-deoxy-D-xylulose-5-phosphate (DXP) pathway and the mevalonic acid (MVA) pathway. *E. coli* is a suitable host for terpenoid production because it contains the DXP pathway genes and can be readily engineered. DXP is initiated by the coupling of G-3-P and pyruvate (Pyr) from glycolysis pathway, while MVA pathway is initiated by condensation of acetyl-CoA.[97] Actually, both DXP and MVA pathways have been harnessed in *E. coli* for the production of valuable terpenoid pharmaceuticals.[98] Optimization of native MEP pathway and taxadiene synthase via multivariate-modular approach elevated the production of taxadiene in *E. coli*, a precursor of anticancer drug Taxol.[99] Expression of MVA pathway genes from *Saccharomyces cerevisiae* and a synthetic amorpha-4,11-diene synthase gene in *E. coli* successfully produced amorphadiene, a precursor of the antimalarial drug artemisinin.[100] Tetraterpene carotenoids such as carotene, lutein, lycopene, and astaxanthin have been widely used in nutraceutical industries as colorants and feed supplements. Carotenoid production in *E. coli* has been established via heterologous expression of carotenoid pathway genes and has been used as an alternative to natural carotenoids.[101–103] To optimize terpenoid synthesis in *E. coli*, multiple attempts have been achieved. Strain optimization via combinatorial gene knockouts increased precursor availability and thus significantly increased lycopene production in *E. coli*.[9,104] Balance of GAP and pyruvate by modulating the reactions catalyzing the conversion of these precursors resulted in an increased production of lycopene.[105] Engineering of central metabolic pathways to increase ATP and NADPH supplies enhanced β-carotene production to 2.1 g/L β-carotene.[106] Metabolic engineering for terpenoid production has shown potential applications for industrial applications because of high value of terpenoids and high efficiency of production.

1.4. Strategies in Metabolic Engineering of Glycolysis

Host strain engineering via gene knockouts of competing pathway or repressors is one of the major strategies in metabolic engineering. Blocking the unnecessary competing genes enhanced the availability of precursor pools and redirect carbon flux to target products via designed pathways. It is easy to manipulate, and the engineered strains are in most cases genetically stable and highly efficient. Metabolic engineering of pyruvate production exemplified the feasibility of systematic gene knockouts for improving pyruvate production and accumulation. Blocking or reducing pyruvate consumption to TCA cycle, acetate and lactate significantly increased the yield of pyruvate from glucose [Figure 1.3(a)].[65] Similarly, the upper intermediate PEP could also be enhanced systematically via genetic modification and reguided for production and aromatic compounds.[60] Especially, the connection between glycolysis pathway with PP pathway, ED pathway, and TCA cycle enabled genetic manipulation to redirect carbon flux to whatever branching node that could be used as the starting material for target chemical production.

However, the gene knockout method holds some inevitable shortcomings like not applicable to essential genes. The development of a general strategy via small RNA (sRNA) mediated interference of target genes is able to modulate expression (especially repression) of any gene at the translation stage using rationally designed synthetic sRNAs.[107,108] Synthetic sRNAs in *E. coli* are composed of a target-binding sequence for target mRNA recognition and a scaffold sequence for recruiting the Hfq protein for mRNA degradation.[15,109] The synthetic sRNA-based gene modulation is advantageous because of its easy implementation and not needing pre-constructed strain mutants. Expression of aromatic pathway genes (*ppsA*, *tktA*, *aroF*, *aroK*, *tyrC*, *aroG^fbr*, and *tyrA^fbr*) and the simultaneous silencing of regulators *tyrR* and *csrA* via counterpart sRNAs in *E. coli* strain S17-1 achieved 21.90 g/L tyrosine production in high-density cultures [Figure 1.3(b)].[15] Interestingly, anti*pgi* sRNA variants finely achieved a balance between biomass formation and tyrosine production.[15] Recently, the antisense RNA (asRNA)-based repression techniques were applied in *E. coli* to directly control glycolytic flux via repression of *glk* to enhance flavonoid production via repression of *fab* operon and to increase polyketide 6-deoxyerythronolide B production via repression of *guaB* and *zwf*.[110–113] Thus, the RNA-based methods are complementary to gene knockout method and possesses some advantages in controlling multigenes including essential genes. The newly developed CRISPR/dCas9 repression system (CRISPRi) holds more promising applications in metabolic engineering because of its higher repression efficacy than sRNA or asRNA method. Clustered regularly interspaced short palindromic repeats (CRISPR)/Cas system is the immune system of bacteria and archaea that is involved in resistance to viruses or plasmids via a small RNA-guided manner.[114–116] CRISPR/Cas9 system from *S. pyogenes* has been developed and proved as a powerful tool for genome editing in either prokaryotic or eukaryotic cells.[117] Under the guidance of synthetic single-guide RNA (sgRNA, ~20 bp) that could target a specific complementary sequence on the genome, an engineered nuclease-deficient Cas9 (dCas9) is directed to bind on target sequence and preclude gene transcription.[118] CRISPRi-mediated metabolic engineering has been recently applied in *E. coli*, for example, to increase polyhydroxyalkanoate (PHA) and (2S)-naringenin production.[119–121] The advantage of CRISPR/dCas9 system lies further in that it can be engineered to realize both repression and activation of genes in both prokaryotic and eukaryotic cells.[122–124]

The roaring development of metabolic engineering enables not only gene regulation, but also protein regulation in a designed manner. Enzymes especially those in glycolysis pathway are sometimes regulated by allosteric regulation like feedback inhibition. Feedback inhibition resistant mutants are always helpful to crease the final product yield, like GlmS variants were resistant to inhibition of glucosamine-6-P and significantly increased glucosamine production.[47] Similar examples are AroG mutant D146N (AroG^{fbr}) and TyrA mutant M53I/A354V (TyrA^{fbr}), which are resistant to inhibition by aromatic amino acids and enhanced production of aromatic compounds. Directed evolution or rational engineering of enzymes is widely used in protein engineering to enhance activity or change substrate specificity. The typical example is the establishment of the citramalate pathway for 1-propanol via directed evolution of citramalate synthase (CimA).[74] Recently, a novel way to dynamically control protein degradation enabled modulation of enzyme activity [Figure 1.3(c)].[40] Controlling the degradation of a key glycolytic enzyme, Pfk I, by fusing it to a degradable signal tag SsrA tag, *E. coli* was rendered in a growth mode close to wild type and an increased G-6-P pool available for *myo*-inositol production.[40]

High-level expression of heterologous genes in *E. coli* may be toxic or redundant to cells while low-level expression may impair productivity. Hence, the most advanced ideal system for metabolic engineering is via dynamic control of metabolic pathways to eliminate metabolic imbalances.[125,126] To achieve this, the commonly used systems are quorum-sensing that modulates gene expression based on cell population, and genetic switches like AND gate or toggle switch that induce gene expression when certain inducers are present.[127] The quorum sensing system has been successfully incorporated into 1,4-butanediol (BD) biosynthesis pathway in *E. coli* and achieved autonomous BD production.[128] Genetic switches are like on–off switches that redistribute carbon flux among various pathways via inducer-mediated control of the key enzymes of the metabolic network.[129] A simple example is a *tetR/lacI*-based genetic switch of citrate synthase (*gltA*) designed for isopropanol production. When IPTG was absent, *gltA* was constitutively expressed; when IPTG was added, *gltA* was repressed and acetyl-CoA was accumulated for isopropanol pathway.[130] The metabolite biosensors was also introduced into dynamic control system that the expression of key pathway enzymes is modulated by the availability of key metabolites or desired products. This creates a cellular production system capable of balancing cellular metabolism to reduce cell stress and increase overall yield. Incorporation of a malonyl-CoA response regulator in *E. coli* is able to dynamically control the expression of critical enzymes involved in the supply and consumption of malonyl-CoA and efficiently redirect carbon flux toward fatty acid biosynthesis.[131] Recently, biosensors for various metabolites or natural products have been revealed, which are potential candidates that could be engineered into systems like monitoring of biosynthesis processes, real-time detection of products, and dynamical control of metabolic pathways [Figure 1.3(d)].[132–135]

1.5. Conclusion and Outlook

Glycolysis pathway is essential for carbon source utilization and cell survival for almost all microorganisms. Glycolysis pathway in *E. coli* has been served as a working horse converting carbon source to metabolite intermediates, which enabled *E. coli* to be a versatile microbial factory for chemical synthesis. Manipulation of glycolysis pathway can convert carbon flux to desired chemicals from important nodes along glycolysis pathway, like from G-6-P to trehalose, and glucaric acid, from F-6-P to glucosamine and polysaccharides, from G-3-P and pyruvate to terpenoids, from PEP to aromatic compounds and from pyruvate to various biofuels. As glycolysis is essential for cell growth, these manipulations are principally focused on balancing the carbon flux for biomass formation and production yield of desired products. Methods like gene knockout, RNA-based repression, CRISPRi, and dynamic control rendered glycolysis pathway to distribute carbon flux in a programmable manner, maximizing the yield of target compounds. However, additional efforts are still needed to increase the productivity and stability of microbial factories for industrial applications.

References

1. Romano, A., & Conway, T. Evolution of carbohydrate metabolic pathways. *Res. Microbiol.* **147**, 448–455 (1996).
2. Nelson, D.L., Lehninger, A.L., & Cox, M.M. *Lehninger Principles of Biochemistry*. Macmillan, New York (2008).
3. Hernández-Montalvo, V. *et al.* Expression of *galP* and *glk* in a *Escherichia coli* PTS mutant restores glucose transport and increases glycolytic flux to fermentation products. *Biotechnol. Bioeng.* **83**, 687–694 (2003).
4. Curtis, S.J., & Epstein, W. Phosphorylation of D-glucose in *Escherichia coli* mutants defective in glucosephosphotransferase, mannosephosphotransferase, and glucokinase. *J. Bacteriol.* **122**, 1189–1199 (1975).
5. Kotlarz, D., Garreau, H., & Buc, H. Regulation of the amount and of the activity of phosphofructokinases and pyruvate kinases in *Escherichia coli*. *Biochim. Biophys. Acta Gen. Subj.* **381**, 257–268 (1975).
6. Marsh, J.J., & Lebherz, H.G. Fructose-bisphosphate aldolases: An evolutionary history. *Trends Biochem. Sci.* **17**, 110–113 (1992).

7. Scamuffa, M.D., & Caprioli, R.M. Comparison of the mechanisms of two distinct aldolases from *Escherichia coli* grown on gluconeogenic substrates. *Biochim. Biophys. Acta Enzymol.* **614**, 583–590 (1980).

8. Macomber, L., Elsey, S.P., & Hausinger, R.P. Fructose-1, 6-bisphosphate aldolase (class II) is the primary site of nickel toxicity in *Escherichia coli*. *Mol. Microbiol.* **82**, 1291–1300 (2011).

9. Alper, H., Jin, Y.-S., Moxley, J., & Stephanopoulos, G. Identifying gene targets for the metabolic engineering of lycopene biosynthesis in *Escherichia coli*. *Metab. Eng.* **7**, 155–164 (2005).

10. Foster, J.M. *et al*. Evolution of bacterial phosphoglycerate mutases: Non-homologous isofunctional enzymes undergoing gene losses, gains and lateral transfers. *PLoS One* **5**, e13576 (2010).

11. Conway, T. The Entner-Doudoroff pathway: History, physiology and molecular biology. *FEMS Microbiol. Rev.* **9**, 1–27 (1992).

12. Flamholz, A., Noor, E., Bar-Even, A., Liebermeister, W., & Milo, R. Glycolytic strategy as a tradeoff between energy yield and protein cost. *Proc. Nat. Acad. Sci.* **110**, 10039–10044 (2013).

13. Plumbridge, J. Regulation of gene expression in the PTS in *Escherichia coli*: The role and interactions of Mlc. *Curr. Opin. Microbiol.* **5**, 187–193 (2002).

14. Morin, M. *et al*. The post-transcriptional regulatory system CSR controls the balance of metabolic pools in upper glycolysis of *Escherichia coli*. *Mol. Microbiol.* **100**(4), 686–700 (2016).

15. Na, D. *et al*. Metabolic engineering of *Escherichia coli* using synthetic small regulatory RNAs. *Nat. Biotechnol.* **31**, 170–174 (2013).

16. Lau, F., & Fersht, A.R. Conversion of allosteric inhibition to activation in phosphofructokinase by protein engineering. *Nature* **326**, 811–812 (1986).

17. Schiraldi, C., Di Lernia, I., & De Rosa, M. Trehalose production: Exploiting novel approaches. *Trends Biotechnol.* **20**, 420–425 (2002).

18. Jung, G.Y., & Stephanopoulos, G. A functional protein chip for pathway optimization and *in vitro* metabolic engineering. *Science* **304**, 428–431 (2004).

19. Li, H. *et al*. Enhanced production of trehalose in *Escherichia coli* by homologous expression of *otsBA* in the presence of the trehalase inhibitor, validamycin A, at high osmolarity. *J. Biosci. Bioeng.* **113**, 224–232 (2012).

20. Strom, A., & Kaasen, I. Trehalose metabolism in *Escherichia coli*: Stress protection and stress regulation of gene expression. *Mol. Microbiol.* **8**, 205–210 (1993).

21. Purvis, J.E., Yomano, L., & Ingram, L. Enhanced trehalose production improves growth of *Escherichia coli* under osmotic stress. *Appl. Environ. Microbiol.* **71**, 3761–3769 (2005).

22. Baumgärtner, F., Seitz, L., Sprenger, G.A., & Albermann, C. Construction of *Escherichia coli* strains with chromosomally integrated expression cassettes for the synthesis of 2′-fucosyllactose. *Microb. Cell Fact.* **12**, 1(2013).

23. Lee, W.-H. *et al.* Whole cell biosynthesis of a functional oligosaccharide, 2′-fucosyllactose, using engineered *Escherichia coli. Microb. Cell Fact.* **11**, 1 (2012).

24. Chin, Y.-W., Kim, J.-Y., Lee, W.-H., & Seo, J.-H. Enhanced production of 2′-fucosyllactose in engineered *Escherichia coli* BL21star (DE3) by modulation of lactose metabolism and fucosyltransferase. *J. Biotechnol.* **210**, 107–115 (2015).

25. Sze, J.H., Brownlie, J.C., & Love, C.A. Biotechnological production of hyaluronic acid: A mini review. *Biotechnology* **6**, 1–9 (2016).

26. Yu, H., & Stephanopoulos, G. Metabolic engineering of *Escherichia coli* for biosynthesis of hyaluronic acid. *Metab. Eng.* **10**, 24–32 (2008).

27. Wu, Q. *et al.* Transcriptional engineering of *Escherichia coli* K4 for fructosylated chondroitin production. *Biotechnol. Prog.* **29**, 1140–1149 (2013).

28. Schiraldi, C., Cimini, D., & De Rosa, M. Production of chondroitin sulfate and chondroitin. *Appl. Microbiol. Biotechnol.* **87**, 1209–1220 (2010).

29. Fu, L., Suflita, M., & Linhardt, R.J. Bioengineered heparins and heparan sulfates. *Adv. Drug Deliv. Rev.* **97**, 237–249 (2016).

30. Suflita, M., Fu, L., He, W., Koffas, M., & Linhardt, R.J. Heparin and related polysaccharides: Synthesis using recombinant enzymes and metabolic engineering. *Appl. Microbiol. Biotechnol.* **99**, 7465–7479 (2015).

31. Zhang, C. *et al.* Metabolic engineering of *Escherichia coli* BL21 for biosynthesis of heparosan, a bioengineered heparin precursor. *Metab. Eng.* **14**, 521–527 (2012).

32. He, W. *et al.* Production of chondroitin in metabolically engineered *E. coli. Metab. Eng.* **27**, 92–100 (2015).

33. Werpy, T. *et al.* (DTIC Document, 2004).

34. Bhattacharya, S., Manna, P., Gachhui, R., & Sil, P.C. D-Saccharic acid 1, 4-lactone protects diabetic rat kidney by ameliorating hyperglycemia-mediated oxidative stress and renal inflammatory cytokines via NF-κB and PKC signaling. *Toxicol. Appl. Pharmacol.* **267**, 16–29 (2013).

35. Shiue, E., & Prather, K.L. Improving D-glucaric acid production from myo-inositol in *E. coli* by increasing MIOX stability and myo-inositol transport. *Metab. Eng.* **22**, 22–31 (2014).

36. Yamaoka, M., Osawa, S., Morinaga, T., Takenaka, S., & Yoshida, K.-I. A cell factory of *Bacillus subtilis* engineered for the simple bioconversion of myo-inositol to scyllo-inositol, a potential therapeutic agent for Alzheimer's disease. *Microb. Cell Fact.* **10**, 1 (2011).

37. Tanaka, K., Tajima, S., Takenaka, S., & Yoshida, K.-I. An improved *Bacillus subtilis* cell factory for producing scyllo-inositol, a promising therapeutic agent for Alzheimer's disease. *Microb. Cell Fact.* **12**, 1 (2013).

38. Moon, T.S., Yoon, S.-H., Lanza, A.M., Roy-Mayhew, J.D., & Prather, K.L.J. Production of glucaric acid from a synthetic pathway in recombinant *Escherichia coli. Appl. Environ. Microbiol.* **75**, 589–595 (2009).

39. Moon, T.S., Dueber, J.E., Shiue, E., & Prather, K.L.J. Use of modular, synthetic scaffolds for improved production of glucaric acid in engineered *E. coli. Metab. Eng.* **12**, 298–305 (2010).

40. Brockman, I.M., & Prather, K.L. Dynamic knockdown of *E. coli* central metabolism for redirecting fluxes of primary metabolites. *Metab. Eng.* **28**, 104–113 (2015).

41. Byrne, J. Glucosamine market reaching maturity. Nutraingredients. com. 15 (2010). http://www.nutraingredients-usa.com/Industry/Glucosamine-market-reaching-maturity.

42. Chen, X., Liu, L., Li, J., Du, G., & Chen, J. Improved glucosamine and N-acetylglucosamine production by an engineered *Escherichia coli* via step-wise regulation of dissolved oxygen level. *Bioresour. Technol.* **110**, 534–538 (2012).

43. Clegg, D.O. *et al.* Glucosamine, chondroitin sulfate, and the two in combination for painful knee osteoarthritis. *New Engl. J. Med.* **354**, 795–808 (2006).

44. Deng, M.-D. *et al.* Metabolic engineering of *Escherichia coli* for industrial production of glucosamine and N-acetylglucosamine. *Metab. Eng.* **7**, 201–214 (2005).

45. Liu, L. *et al.* Microbial production of glucosamine and N-acetylglucosamine: Advances and perspectives. *Appl. Microbiol. Biotechnol.* **97**, 6149–6158 (2013).

46. Deng, M.-D., Wassink, S.L., & Grund, A.D. Engineering a new pathway for N-acetylglucosamine production: Coupling a catabolic enzyme, glucosamine-6-phosphate deaminase, with a biosynthetic enzyme, glucosamine-6-phosphate N-acetyltransferase. *Enzyme. Microb. Technol.* **39**, 828–834 (2006).

47. Deng, M.-D. *et al.* Directed evolution and characterization of *Escherichia coli* glucosamine synthase. *Biochimie* **88**, 419–429 (2006).
48. Chen, X. *et al.* Optimization of glucose feeding approaches for enhanced glucosamine and N-acetylglucosamine production by an engineered *Escherichia coli. J. Ind. Microbiol. Biotechnol.* **39**, 359–365 (2012).
49. Lin, Y., Sun, X., Yuan, Q., & Yan, Y. Engineering bacterial phenylalanine 4-hydroxylase for microbial synthesis of human neurotransmitter precursor 5-hydroxytryptophan. *ACS Synth. Biol.* **3**, 497–505 (2014).
50. Rajan, P. *et al.* Synthesis and evaluation of caffeic acid amides as antioxidants. *Bioorg. Med. Chem. Lett.* **11**, 215–217 (2001).
51. Touaibia, M., Jean-Francois, J., & Doiron, J. Caffeic acid, a versatile pharmacophore: An overview. *Mini Rev. Med. Chem.* **11**, 695–713 (2011).
52. Sova, M. Antioxidant and antimicrobial activities of cinnamic acid derivatives. *Mini Rev. Med. Chem.* **12**, 749–767 (2012).
53. Hua, D., & Xu, P. Recent advances in biotechnological production of 2-phenylethanol. *Biotechnol. Adv.* **29**, 654–660 (2011).
54. Lin, Y., & Yan, Y. Biosynthesis of caffeic acid in *Escherichia coli* using its endogenous hydroxylase complex. *Microb. Cell Fact.* **11**, 1 (2012).
55. Lin, Y., Jain, R., & Yan, Y. Microbial production of antioxidant food ingredients via metabolic engineering. *Curr. Opin. Biotechnol.* **26**, 71–78 (2014).
56. Wang, J., Guleria, S., Koffas, M.A., & Yan, Y. Microbial production of value-added nutraceuticals. *Curr. Opin. Biotechnol.* **37**, 97–104 (2016).
57. Gosset, G. Improvement of *Escherichia coli* production strains by modification of the phosphoenolpyruvate: Sugar phosphotransferase system. *Microb. Cell Fact.* **4**, 1 (2005).
58. Flores, N., Xiao, J., Berry, A., Bolivar, F., & Valle, F. Pathway engineering for the production of aromatic compounds in *Escherichia coli. Nat. Biotechnol.* **14**, 620–623 (1996).
59. Yi, J., Draths, K., Li, K., & Frost, J. Altered glucose transport and shikimate pathway product yields in *E. coli. Biotechnol. Prog.* **19**, 1450–1459 (2003).
60. Lütke-Eversloh, T., & Stephanopoulos, G. L-tyrosine production by deregulated strains of *Escherichia coli. Appl. Microbiol. Biotechnol.* **75**, 103–110 (2007).
61. Bulter, T., Bernstein, J.R., & Liao, J.C. A perspective of metabolic engineering strategies: Moving up the systems hierarchy. *Biotechnol. Bioeng.* **84**, 815–821 (2003).
62. Rodriguez, A. *et al.* Engineering *Escherichia coli* to overproduce aromatic amino acids and derived compounds. *Microb. Cell Fact.* **13**, 1 (2014).

63. Shimizu, K. Regulation systems of bacteria such as *Escherichia coli* in response to nutrient limitation and environmental stresses. *Metabolites* **4**, 1–35 (2013).

64. Zhu, Y., Eiteman, M.A., Altman, R., & Altman, E. High glycolytic flux improves pyruvate production by a metabolically engineered *Escherichia coli* strain. *Appl. Environ. Microbiol.* **74**, 6649–6655 (2008).

65. Causey, T., Shanmugam, K., Yomano, L., & Ingram, L. Engineering *Escherichia coli* for efficient conversion of glucose to pyruvate. *Proc. Natl. Acad. Sci. USA* **101**, 2235–2240 (2004).

66. Zhu, Y., Eiteman, M., DeWitt, K., & Altman, E. Homolactate fermentation by metabolically engineered *Escherichia coli* strains. *Appl. Environ. Microbiol.* **73**, 456–464 (2007).

67. Stephanopoulos, G. Challenges in engineering microbes for biofuels production. *Science* **315**, 801–804 (2007).

68. Saini, J.K., Saini, R., & Tewari, L. Lignocellulosic agriculture wastes as biomass feedstocks for second-generation bioethanol production: Concepts and recent developments. *Biotech.* **5**, 337–353 (2015).

69. Lindsay, S., Bothast, R., & Ingram, L. Improved strains of recombinant *Escherichia coli* for ethanol production from sugar mixtures. *Appl. Microbiol. Biotechnol.* **43**, 70–75 (1995).

70. Ingram, L., Conway, T., Clark, D., Sewell, G., & Preston, J. Genetic engineering of ethanol production in *Escherichia coli*. *Appl. Environ. Microbiol.* **53**, 2420–2425 (1987).

71. Yomano, L., York, S., Zhou, S., Shanmugam, K., & Ingram, L. Re-engineering *Escherichia coli* for ethanol production. *Biotechnol. Lett.* **30**, 2097–2103 (2008).

72. Trinh, C.T., Unrean, P., & Srienc, F. Minimal *Escherichia coli* cell for the most efficient production of ethanol from hexoses and pentoses. *Appl. Environ. Microbiol.* **74**, 3634–3643 (2008).

73. Atsumi, S., Hanai, T., & Liao, J.C. Non-fermentative pathways for synthesis of branched-chain higher alcohols as biofuels. *Nature* **451**, 86–89 (2008).

74. Atsumi, S., & Liao, J.C. Directed evolution of Methanococcus jannaschii citramalate synthase for biosynthesis of 1-propanol and 1-butanol by *Escherichia coli*. *Appl. Environ. Microbiol.* **74**, 7802–7808 (2008).

75. Jain, R., & Yan, Y. Dehydratase mediated 1-propanol production in metabolically engineered *Escherichia coli*. *Microb. Cell Fact.* **10**, 1 (2011).

76. Shen, C.R., & Liao, J.C. Metabolic engineering of *Escherichia coli* for 1-butanol and 1-propanol production via the keto-acid pathways. *Metab. Eng.* **10**, 312–320 (2008).

77. Shen, C.R., & Liao, J.C. Synergy as design principle for metabolic engineering of 1-propanol production in *Escherichia coli*. *Metab. Eng.* **17**, 12–22 (2013).

78. Atsumi, S. *et al*. Metabolic engineering of *Escherichia coli* for 1-butanol production. *Metab. Eng.* **10**, 305–311 (2008).

79. Smith, K.M., & Liao, J.C. An evolutionary strategy for isobutanol production strain development in *Escherichia coli*. *Metab. Eng.* **13**, 674–681 (2011).

80. Shi, A., Zhu, X., Lu, J., Zhang, X., & Ma, Y. Activating transhydrogenase and NAD kinase in combination for improving isobutanol production. *Metab. Eng.* **16**, 1–10 (2013).

81. Connor, M.R., Cann, A.F., & Liao, J.C. 3-Methyl-1-butanol production in *Escherichia coli*: Random mutagenesis and two-phase fermentation. *Appl. Microbiol. Biotechnol.* **86**, 1155–1164 (2010).

82. Show, K., Lee, D., Tay, J., Lin, C., & Chang, J. Biohydrogen production: Current perspectives and the way forward. *Int. J. Hydrogen Energy.* **37**, 15616–15631 (2012).

83. Cai, G., Jin, B., Monis, P., & Saint, C. Metabolic flux network and analysis of fermentative hydrogen production. *Biotechnol. Adv.* **29**, 375–387 (2011).

84. Yoshida, A., Nishimura, T., Kawaguchi, H., Inui, M., & Yukawa, H. Enhanced hydrogen production from formic acid by formate hydrogen lyase-overexpressing *Escherichia coli* strains. *Appl. Environ. Microbiol.* **71**, 6762–6768 (2005).

85. Sawers, R. Formate and its role in hydrogen production in *Escherichia coli*. *Biochem. Soc. Trans.* **33**, 42–46 (2005).

86. Maeda, T., Sanchez-Torres, V., & Wood, T.K. Hydrogen production by recombinant *Escherichia coli* strains. *Microb. Biotechnol.* **5**, 214–225 (2012).

87. Rosales-Colunga, L.M., & Rodríguez, A.D.L. *Escherichia coli* and its application to biohydrogen production. *Rev. Environ. Sci. Biotechnol.* **14**, 123–135 (2015).

88. Lee, H.-S., Vermaas, W.F., & Rittmann, B.E. Biological hydrogen production: Prospects and challenges. *Trends Biotechnol.* **28**, 262–271 (2010).

89. Kim, S., Seol, E., Oh, Y.-K., Wang, G., & Park, S. Hydrogen production and metabolic flux analysis of metabolically engineered *Escherichia coli* strains. *Int. J. Hydrogen Energy.* **34**, 7417–7427 (2009).

90. Veit, A., Akhtar, M.K., Mizutani, T., & Jones, P.R. Constructing and testing the thermodynamic limits of synthetic NAD (P) H: H_2 pathways. *Microb. Biotechnol.* **1**, 382–394 (2008).

91. Tran, K.T., Maeda, T., & Wood, T.K. Metabolic engineering of *Escherichia coli* to enhance hydrogen production from glycerol. *Appl. Microbiol. Biotechnol.* **98**, 4757–4770 (2014).

92. Sekar, B.S., Seol, E., Raj, S.M., & Park, S. Co-production of hydrogen and ethanol by *pfkA*-deficient *Escherichia coli* with activated pentose-phosphate pathway: Reduction of pyruvate accumulation. *Biotechnol. Biofuels* **9**, 1 (2016).

93. Jain, N., & Ramawat, K.G. *Natural products.* Springer, Heidelberg, Berlin, 2559–2580 (2013).

94. Mora-Pale, M., Sanchez-Rodriguez, S.P., Linhardt, R.J., Dordick, J.S., & Koffas, M.A. Metabolic engineering and *in vitro* biosynthesis of phytochemicals and non-natural analogues. *Plant Sci.* **210**, 10–24 (2013).

95. Netzer, R. *et al.* Biosynthetic pathway for γ-cyclic sarcinaxanthin in *Micrococcus luteus*: Heterologous expression and evidence for diverse and multiple catalytic functions of C50 carotenoid cyclases. *J Bacteriol.* **192**, 5688–5699 (2010).

96. Furubayashi, M. *et al.* A highly selective biosynthetic pathway to non-natural C50 carotenoids assembled from moderately selective enzymes. *Nat. Commun.* **6**, 7534 (2015).

97. Vranová, E., Coman, D., & Gruissem, W. Network analysis of the MVA and MEP pathways for isoprenoid synthesis. *Annu. Rev. Plant Biol.* **64**, 665–700 (2013).

98. Morrone, D. *et al.* Increasing diterpene yield with a modular metabolic engineering system in *E. coli*: Comparison of MEV and MEP isoprenoid precursor pathway engineering. *Appl. Microbiol. Biotechnol.* **85**, 1893–1906 (2010).

99. Ajikumar, P.K. *et al.* Isoprenoid pathway optimization for Taxol precursor overproduction in *Escherichia coli*. *Science* **330**, 70–74 (2010).

100. Martin, V.J., Pitera, D.J., Withers, S.T., Newman, J.D., & Keasling, J.D. Engineering a mevalonate pathway in *Escherichia coli* for production of terpenoids. *Nat. Biotechnol.* **21**, 796–802 (2003).

101. Albrecht, M., Misawa, N., & Sandmann, G. Metabolic engineering of the terpenoid biosynthetic pathway of *Escherichia coli* for production of the carotenoids β-carotene and zeaxanthin. *Biotechnol. Lett.* **21**, 791–795 (1999).

102. Matthews, P., & Wurtzel, D.E.T. Metabolic engineering of carotenoid accumulation in *Escherichia coli* by modulation of the isoprenoid precursor pool with expression of deoxyxylulose phosphate synthase. *Appl. Microbiol. Biotechnol.* **53**, 396–400 (2000).

103. Kim, S.W., & Keasling, J. Metabolic engineering of the nonmevalonate isopentenyl diphosphate synthesis pathway in *Escherichia coli* enhances lycopene production. *Biotechnol. Bioeng.* **72**, 408–415 (2001).

104. Alper, H., Miyaoku, K., & Stephanopoulos, G. Construction of lycopene-overproducing *E. coli* strains by combining systematic and combinatorial gene knockout targets. *Nat. Biotechnol.* **23**, 612–616 (2005).

105. Farmer, W.R., & Liao, J.C. Precursor balancing for metabolic engineering of lycopene production in *Escherichia coli*. *Biotechnol. Prog.* **17**, 57–61 (2001).

106. Zhao, J. *et al.* Engineering central metabolic modules of *Escherichia coli* for improving β-carotene production. *Metab. Eng.* **17**, 42–50 (2013).

107. Gottesman, S. The small RNA regulators of *Escherichia coli*: Roles and mechanisms. *Annu. Rev. Microbiol.* **58**, 303–328 (2004).

108. Aiba, H. Mechanism of RNA silencing by Hfq-binding small RNAs. *Curr. Opin. Microbiol.* **10**, 134–139 (2007).

109. Pichon, C., Du Merle, L., Caliot, M.E., Trieu-Cuot, P., & Le Bouguénec, C. An *in silico* model for identification of small RNAs in whole bacterial genomes: Characterization of antisense RNAs in pathogenic *Escherichia coli* and Streptococcus agalactiae strains. *Nucleic Acids Res.* **40**, 2846–2861 (2012).

110. Solomon, K.V., Sanders, T.M., & Prather, K.L. A dynamic metabolite valve for the control of central carbon metabolism. *Metab. Eng.* **14**, 661–671 (2012).

111. Yang, Y., Lin, Y., Li, L., Linhardt, R.J., & Yan, Y. Regulating malonyl-CoA metabolism via synthetic antisense RNAs for enhanced biosynthesis of natural products. *Metab. Eng.* **29**, 217–226 (2015).

112. Wu, J., Yu, O., Du, G., Zhou, J., & Chen, J. Fine-tuning of the fatty acid pathway by synthetic antisense RNA for enhanced (2S)-naringenin production from L-tyrosine in *Escherichia coli*. *Appl. Environ. Microbiol.* **80**, 7283–7292 (2014).

113. Meng, H.L., Xiong, Z.Q., Song, S.J., Wang, J., & Wang, Y. Construction of polyketide overproducing *Escherichia coli* strains via synthetic antisense RNAs based on *in silico* fluxome analysis and comparative transcriptome analysis. *Biotechnol. J.* **11**(4), 530–541 (2016).

114. Barrangou, R. *et al.* CRISPR provides acquired resistance against viruses in prokaryotes. *Science* **315**, 1709–1712 (2007).

115. Brouns, S.J. *et al.* Small CRISPR RNAs guide antiviral defense in prokaryotes. *Science* **321**, 960–964 (2008).

116. Horvath, P., & Barrangou, R. CRISPR/Cas, the immune system of bacteria and archaea. *Science* **327**, 167–170 (2010).

117. Cong, L. *et al.* Multiplex genome engineering using CRISPR/Cas systems. *Science* **339**, 819–823 (2013).

118. Qi, L.S. *et al.* Repurposing CRISPR as an RNA-guided platform for sequence-specific control of gene expression. *Cell* **152**, 1173–1183 (2013).

119. Jakočiūnas, T., Jensen, M.K., & Keasling, J.D. CRISPR/Cas9 advances engineering of microbial cell factories. *Metab. Eng.* **34**, 44–59 (2016).

120. Wu, J., Du, G., Chen, J., & Zhou, J. Enhancing flavonoid production by systematically tuning the central metabolic pathways based on a CRISPR interference system in *Escherichia coli. Sci. Rep.* **5**, 13477 (2015), doi: 10.1038/srep13477.

121. Lv, L., Ren, Y.-L., Chen, J.-C., Wu, Q., & Chen, G.-Q. Application of CRISPRi for prokaryotic metabolic engineering involving multiple genes, a case study: Controllable P (3HB-co-4HB) biosynthesis. *Metab. Eng.* **29**, 160–168 (2015).

122. Perez-Pinera, P. *et al.* RNA-guided gene activation by CRISPR-Cas9-based transcription factors. *Nat. Methods* **10**, 973–976 (2013).

123. Bikard, D. *et al.* Programmable repression and activation of bacterial gene expression using an engineered CRISPR-Cas system. *Nucleic Acids Res.* **41**, 7429–7437 (2013).

124. Zalatan, J.G. *et al.* Engineering complex synthetic transcriptional programs with CRISPR RNA scaffolds. *Cell* **160**, 339–350 (2015).

125. Zhang, F., Carothers, J.M., & Keasling, J.D. Design of a dynamic sensor-regulator system for production of chemicals and fuels derived from fatty acids. *Nat. Biotechnol.* **30**, 354–359 (2012).

126. Venayak, N., Anesiadis, N., Cluett, W.R., & Mahadevan, R. Engineering metabolism through dynamic control. *Curr. Opin. Biotechnol.* **34**, 142–152 (2015).

127. Tsao, C.-Y., Hooshangi, S., Wu, H.-C., Valdes, J.J., & Bentley, W.E. Autonomous induction of recombinant proteins by minimally rewiring native quorum sensing regulon of *E. coli. Metab. Eng.* **12**, 291–297 (2010).

128. Liu, H., & Lu, T. Autonomous production of 1, 4-butanediol via a de novo biosynthesis pathway in engineered *Escherichia coli. Metab. Eng.* **29**, 135–141 (2015).

129. Stephanopoulos, G. Synthetic biology and metabolic engineering. *ACS Synth. Biol.* **1**, 514–525 (2012).

130. Soma, Y., Tsuruno, K., Wada, M., Yokota, A., & Hanai, T. Metabolic flux redirection from a central metabolic pathway toward a synthetic pathway using a metabolic toggle switch. *Metab. Eng.* **23**, 175–184 (2014).
131. Xu, P., Li, L., Zhang, F., Stephanopoulos, G., & Koffas, M. Improving fatty acids production by engineering dynamic pathway regulation and metabolic control. *Proc. Natl. Acad. Sci.* **111**, 11299–11304 (2014).
132. Zhang, F., & Keasling, J. Biosensors and their applications in microbial metabolic engineering. *Trends Microbiol.* **19**, 323–329 (2011).
133. Zhang, J., Jensen, M.K., & Keasling, J.D. Development of biosensors and their application in metabolic engineering. *Curr. Opin. Chem. Biol.* **28**, 1–8 (2015).
134. Siedler, S., Stahlhut, S.G., Malla, S., Maury, J., & Neves, A.R. Novel biosensors based on flavonoid-responsive transcriptional regulators introduced into *Escherichia coli*. *Metab. Eng.* **21**, 2–8 (2014).
135. Rogers, J.K. *et al.* Synthetic biosensors for precise gene control and real-time monitoring of metabolites. *Nucleic Acids Res.* **43**(15), 7648–7660 (2015).

Chapter 2

Citric Acid Cycle and Its Metabolic Engineering Applications

Jia Wang[*,†] *and Xiaolin Shen*[*,†,‡]

[*] *State Key Laboratory of Chemical Resource Engineering*
Beijing University of Chemical Technology
Beijing 100029, China
[†] *Beijing Advanced Innovation Center for Soft Matter Science*
and Engineering, Beijing University of Chemical Technology
Beijing 100029, China
[‡] *shenxl@mail.buct.edu.cn*

2.1. Introduction

Organic acids, amino acids and their derived compounds, such as succinic acid, citric acid, glutamic acid, 1,4-butanediol, glutaconic acid, and adipic acid, are widely used in clinical, cosmetic, pharmaceutical, and chemical industry.[1-5] These compounds continued to be produced from petroleum since the past century. However, with reduction in the production of petroleum due to its depleting reserves, researchers are engaged in attempts to produce these compounds by using metabolic engineering approaches. Fortunately, citric acid cycle provides energy and skeletal structure of molecules for life activities. Therefore, most organic acids are intermediates of the citric acid cycle in microorganisms. At the same time, citric acid

35

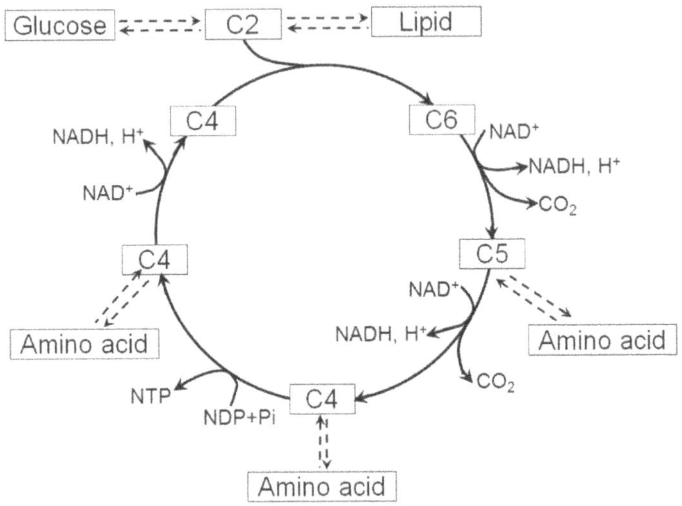

Figure 2.1.   The schematic figure of citric acid cycle. This figure shows the changes in the number of carbon atoms in one cycle. Citric acid cycle is a metabolic hub of sugar, lipid, and amino acid in organisms. This cycle provides various skeletal structure of molecules for life.

cycle is a common metabolic hub of sugar, lipid, and amino acid. This situation gives us more opportunity to modify or reconstruct the metabolic pathways to accumulate the target products (Figure 2.1). Citric acid cycle is also known as tricarboxylic acid (TCA) cycle because the first acid in this cycle has three carboxyl groups. In 1953, Hans Adolf Krebs won the Nobel Price for clarifying the mechanism of citric acid cycle. In memory of Hans Adolf Krebs, this cycle is called Krebs cycle as well. In this chapter, we will introduce the fundamentals of citric acid cycle and its metabolic engineering applications in recent years.

2.2. Preparation of Citric Acid Cycle

Glucose is converted into pyruvate via glycolysis. Then pyruvate reacts through two different pathways depending on the conditions. Under anaerobic conditions, pyruvate generates either lactic acid or ethanol in different organisms. Under aerobic conditions, pyruvate is converted into acetyl coenzyme A (acetyl-CoA) and subsequently decomposes into water and carbon dioxide by oxidation.

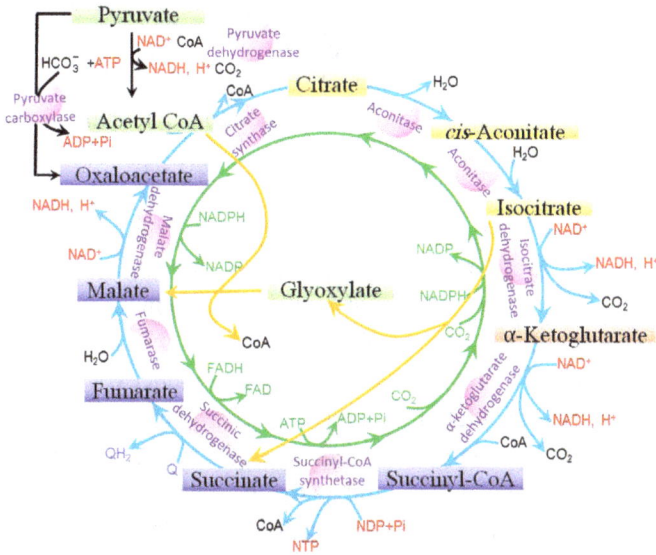

Figure 2.2. Citric acid cycle, reverse citric acid cycle, and glyoxylate cycle. The black arrows show the preparation of citric acid cycle. The blue lines indicate citric acid cycle. The yellow arrows represent glyoxylate cycle. The purple texts indicate the enzymes involved in citric acid cycle. The red texts show the energy-rich molecules.

As shown in Figure 2.2, pyruvate is converted into acetyl-CoA with pyruvate dehydrogenase complex acting as a catalyst. This enzyme complex is a highly ordered and integrated containing three enzymes — pyruvate dehydrogenase component, dihydrolipoyl transacetylase, and dihydrolipoyl dehydrogenase — and uses thiamine pyrophosphate (TPP), lipoamide, CoA, NAD^+, and FAD as cofactors. Each enzyme in pyruvate dehydrogenase complex has its independent active site that can catalyze reactions sequentially. By the catalysis of these three enzymes, pyruvate generates one molecular carbon dioxide and reassigns one high-transfer-potential electron to NAD^+. This reaction is irreversible and connects glycolysis and the citric acid cycle. On the one hand, carbohydrate can be totally oxidized to carbon dioxide and NADH is formed to store energy through this pathway. On the other hand, carbohydrate is converted into fatty acids by this reaction. It means that pyruvate can be broken down to energy or energetic metabolites.

2.3. Fundamentals of Citric Acid Cycle

There are three important functions of the citric acid cycle: oxidization of carbon atoms, obtaining the molecular skeleton, and harvest of energy (Figure 2.1). However, citric acid cycle only directly produces one molecular ATP or GTP in some organisms and does not directly use oxygen in the reactions. How does it work to achieve these goals?

In citric acid cycle, a four-carbon oxaloacetic acid condenses with a two-carbon acetyl-CoA to produce one six-carbon citric acid, which enters the citric acid cycle. Then the six-carbon citric acid loses two-carbon atoms sequentially via oxidation and produces a series of four-carbon organic acids, including succinic acid, fumaric acid, and malic acid. Finally, oxaloacetic acid is regenerated and goes into next cycle. Citric acid cycle is an amphibolic pathway having decomposition and synthesis functions. Based on this, citric acid cycle can be divided into two stages. One is oxidation of two-carbon atoms to collect high-energy electrons and the other is regeneration of oxaloacetic acid to harvest energy-rich electrons. This process provides different number of carbon intermediates as the molecular skeleton to participate in the metabolic process. Those intermediates are the bio-based organic acids or the precursors of chemical products (Figure 2.1).

2.3.1. *Oxidation of Two-Carbon Atoms to Produce High-Energy Electrons*

In the first stage of the citric acid cycle, oxaloacetic acid condenses with acetyl-CoA to generate citric acid, thus achieving the conversion of a four-carbon molecule to a six-carbon molecule (Figure 2.1). Through oxidative decarboxylation, two-carbon atoms are lost in the form of carbon dioxide and coupled with the generation of one four-carbon acid and high-energy electrons.

2.3.1.1. *Oxidation*

Citrate synthase catalyzes the aldol Claisen ester condensation of oxaloacetic acid, acetyl-CoA, and H_2O to produce citric acid and CoA.

Citrate synthase is a rate-limiting enzyme that can be regulated by ATP, NADH, and succinyl-CoA (Figure 2.2). Acetonyl-CoA is an inhibitor of this enzyme. This mechanism is always used to control the carbon flux and enhance the yield of products in the citric acid cycle. Because the primary target of citric acid cycle is the oxidation of carbon atoms to generate high-energy electrons, citric acid should be converted into isocitric acid, which is an optimal form for oxidation. This reaction is catalyzed by aconitase as *cis*-aconitic acid is an intermediate. Through dehydration and hydration, one H atom and one OH group of citric acid have been exchanged to form isocitric acid.

2.3.1.2. *Oxidation and decarboxylation*

The first oxidative decarboxylation of four oxidation–reduction reactions in citric acid cycle is the conversion of isocitric acid to α-ketoglutaric acid. The intermediate of this reaction is oxalosuccinate, which is an unstable compound. Isocitrate dehydrogenase catalyzes this oxidative decarboxylation, leading to the loss of one molecular carbon dioxide and yields NADH, the first high-energy electron carrier, as well as α-ketoglutaric acid (Figure 2.2). This reaction directly leads to a six-carbon compound converted into a five-carbon compound.

After isocitric acid has been decarboxylated, the second oxidative decarboxylation reaction occurs. A carbon atom is lost from α-ketoglutaric acid, leading to the formation of the four-carbon succinyl-CoA and one molecule of NADH (Figure 2.2). This reaction is irreversible and stores energy in the form of the thioester bond in succinyl-CoA. An enzyme complex named α-ketoglutarate dehydrogenase (KGD) complex catalyzes this reaction. Like pyruvate dehydrogenase complex, KGD complex contains three subunits: α-ketoglutarate dehydrogenase, dihydrolipoyl *trans*-succinylase, and dihydrolipoyl dehydrogenase. The catalytic mechanism of this enzyme is the same as pyruvate dehydrogenase complex that uses TPP, lipoic acid, CoA, FAD, NAD^+, and Mg^{2+} as cofactors. Until now, the two-carbon atoms that were added to citric acid are consumed and yield two molecules of NADH. Stage one is finished and the cycle goes into four-carbon organic acid transfer stage.

2.3.2. *Regeneration of Oxaloacetic Acid to Harvest Energy-Rich Electrons*

This stage of citric acid cycle involves several four-carbon organic acids to regenerate the initial compound of this cycle, oxaloacetic acid, and harvest high energy.

2.3.2.1. *Substrate-level phosphorylation*

Compared with ATP, which has a $\Delta G^{0\prime}$ of −30.5 kJ mol^{-1} (−7.3 kcal mol^{-1}), succinyl-CoA is a high-energy thioester molecule having a $\Delta G^{0\prime}$ value of −33.5 kJ mol^{-1} (−8.0 kcal mol^{-1}). The breakdown of this energy-rich thioester bond is coupled with the phosphorylation of a purine nucleoside diphosphate (NDP). In mammal cells, the purine NDP is GDP or ADP depending on the different tissues, but plant cells or microorganisms usually use ADP as the acceptor. It's worth noting that this is the only substrate-level phosphorylation in citric acid cycle, so this reaction directly generates one molecule of GTP/ATP as well as succinic acid. Succinyl-CoA synthetase (SUCLG), also named succinyl thiokinase, catalyzes this reversible reaction. First, CoA group is replaced by an *ortho*-phosphate group to yield succinyl phosphate. In this process, one energy-rich molecule is generated. Then, the histidine residue, which exists in the activity center as the intermediate, transfers the phosphate group to NDP to yield the high-energy compound, nucleoside triphosphate (NTP).

2.3.2.2. *Regeneration of oxaloacetic acid via oxidization of succinic acid*

Oxaloacetic acid is regenerated from succinic acid in three steps: oxidation, reduction, and oxidation occurring sequentially. The new oxaloacetic acid will participate in the next cycle. The first step in this part is the oxidation of succinic acid to generate fumaric acid (Figure 2.2). FAD as the cofactor accepts two hydrogens and is converted into FADH$_2$. The enzyme catalyzing this reaction is different from the other enzymes in the citric acid cycle. This enzyme named succinate dehydrogenase (SDH) is located in the inner mitochondrial membrane in eukaryotes. Because the electron-transport chain is also embedded in the inner mitochondrial membrane, this situation makes SDH directly linked with the electron-transport chain. That means the electron-transport chain is a bridge between the citric acid cycle and the ATP generation system.

Following oxidation is the hydration of fumaric acid to generate L-malic acid. Fumarase catalyzes the breakdown of the carbon–carbon double bond of fumaric acid, leading to the formation of a carbon–hydrogen bond and a carbon–hydroxyl bond by adding an H_2O molecule (Figure 2.2).

The last step of the citric acid cycle is the oxidation of L-malic acid, which is catalyzed by malate dehydrogenase to form oxaloacetic acid. In this reaction, NAD^+ as the hydrogen acceptor is reduced to an energy-rich molecule, NADH (Figure 2.2). NADH enters the electron-transport chain and oxaloacetic acid goes into the next cycle.

2.4. Regulation of Citric Acid Cycle

The citric acid cycle provides energy for cells as well as the building block molecule. It is the common metabolic hub of sugar, lipid, and amino acid. Therefore, it should be subjected to complex and fine regulation.

Although there are nine enzymes/enzyme complexes participating in citric acid cycle, only three of them are important control points under regulation. Those are citrate synthase, isocitric acid dehydrogenase, and KGD complex. Citrate synthase is regulated by the concentration of acetyl-CoA and/or oxaloacetic acid. The concentration of acetyl-CoA is determined by the activity of pyruvate dehydrogenase complex, and the concentration of oxaloacetic acid is influenced by L-malic acid and the ratio of $[NADH]/[NAD^+]$. When the cell consumes a lot of energy, the concentration of NADH is decreased to up-regulate the activity of citrate synthase. Isocitric acid dehydrogenase is the second control point inhibited by NADH. ADP, the signal of lack of energy, can improve the activity of isocitric acid dehydrogenase. Because KGD complex contains three enzymes, this complex is inhibited by all the products of these enzymes. The speed of the cycle is lowered when the concentration of ATP and NADH is high. Additionally, Ca^{2+} can activate isocitric acid dehydrogenase and KGD complex.

Those regulations not only help cells maintain the best situation, but also give us a guide to engineer citric acid cycle in metabolic engineering.

2.5. Citric Acid Cycle-Related Pathways

Two kinds of cycles are based on citric acid cycle to provide organic acids for life activity in some microorganisms and plants under abnormal conditions such as lack of oxygen, denutrition, germination stage, and so on. The first one is the reverse citric acid cycle. This cycle shares almost the same route with citric acid cycle but uses carbon dioxide and H_2O to synthesize organic molecules (Figure 2.2). This cycle plays a pivotal role in bio-organic acid production under oxygen-limited condition in metabolic engineering. The other one is glyoxylate cycle. A part of the citric acid cycle compounds are involved in this cycle. Isocitric acid breaks down to form glyoxylate and succinic acid with isocitrate lyase as the catalyst. Then, glyoxylate condenses with acetyl-CoA to generate malic acid and CoA using malate synthase (Figure 2.2). This cycle is always used to accumulate succinic acid, fumaric acid, and malic acid in the metabolic engineering approach.

Considering that the molecules in the citric acid cycle should produce amino acids to participate in life activities, four anaplerotic reactions of citric acid cycle exist in organisms to maintain the molecular balance of this cycle. The first reaction is the conversion of pyruvate to oxaloacetic acid, which is catalyzed by pyruvate carboxylase. The enhancement of acetyl-CoA's concentration indicates the lack of oxaloacetic acid, so this enzyme is activated by acetyl-CoA. Moreover, pyruvate can be generated from malic acid in a similar reaction. The second method uses amino acids to generate organic acids in citric acid cycle, such as by conversion of aspartic acid to oxaloacetic acid and glutamic acid to α-ketoglutaric acid. The third one is the malonyl-CoA from β-oxidation of odd-chain fatty acids to form succinyl-CoA. The last one is the conversion of adenylosuccinate to fumaric acid.

As mentioned above, the citric acid cycle can provide not only bio-based organic acids but also the precursors of important chemicals. In recent years, more and more bulk or fine chemicals are produced by

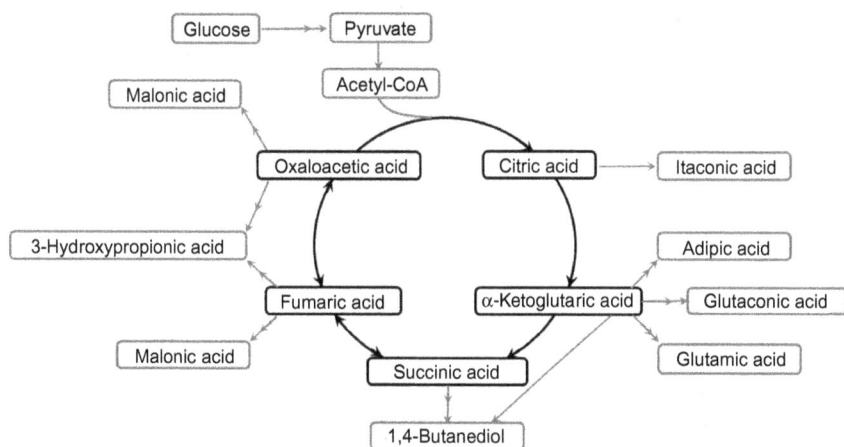

Figure 2.3. Citric acid cycle intermediates and their derivatives.

engineering the citric acid cycle in microorganisms. Then, we review the recent metabolic engineering applications for products from the citric acid cycle (Figure 2.3 and Table 2.1).

2.6. Citric Acid and Its Derived Compounds

2.6.1. *Citric Acid*

Citric acid as an intermediate in the citric acid cycle, synthesized by condensation of oxaloacetate with acetyl-CoA via citrate synthase. It is widely used in food and beverage industry owing to its edibility and pleasant acid taste. In addition, citric acid has been also utilized in pharmaceuticals, cosmetics, detergents, and a range of other industrial applications because of its high water solubility, chelating ability, and buffering property.[2] Up to two million tons of citric acid are produced every year and its market size is still growing.[6]

Currently, industrial production of citric acid relies on fermentation, mainly produced through submerged fermentation using starch or sucrose as the carbon source by *Aspergillus niger* due to its ability to utilize various of raw materials and its high citric acid yield at lower pH value.[7] In 1913, Zahorsky first reported that *A. niger* can overproduce

Table 2.1. A summary of the literature on the production of citric acid cycle-based chemicals.

Product	Titer (g/L)	Yield (g/g)	Carbon source	Host	References
Citric acid	55	0.64	Glucose	*Aspergillus niger*	[9]
	46	—	Glucose	*Aspergillus niger*	[10]
	151 and 135	—	Liquefied corn	*Aspergillus niger*	[13]
	114	0.76	Cane molasse	*Aspergillus niger*	[80]
	43	0.56	Stearin and glucose	*Yarrowia lipolytica*	[16]
	140	0.82	Sucrose	*Yarrowia lipolytica*	[17]
	84	—	Inulin	*Yarrowia lipolytica*	[18]
Itaconic acid	86	0.62	Glucose	*Aspergillus terreus*	[21]
	129	0.51	Glucose	*Aspergillus terreus*	[22]
	18	—	Corn cob, cotton stalk, and sunflower stalk	*Aspergillus terreus*	[24]
	4.3	0.15	Glucose	*Escherichia coli*	[25]
	32	0.49	Glucose	*Escherichia coli*	[26]
α-Ketoglutaric acid	195	—	*n*-Paraffin	*Yarrowia lipolytica*	[31]
	66.2	—	Glycerol	*Yarrowia lipolytica*	[32]
	62.5	—	Glycerol	*Yarrowia lipolytica*	[30]
	186	0.36	Glycerol	*Yarrowia lipolytica*	[27]
	132.7	—	Glycerol	*Yarrowia lipolytica*	[33]

Product			Substrate	Organism	Ref.
	56.5	0.57	Glycerol	*Yarrowia lipolytica*	[34]
	43.3	—	Glycerol	*Yarrowia lipolytica*	[35]
L-Glutamate	32	—	Glucose	*Corynebacterium glutamicum*	[38]
	37	—	Glucose	*Corynebacterium glutamicum*	[41]
	40.5	—	Glucose	*Corynebacterium glutamicum*	[3]
	12.3	0.69	Glucose	*Corynebacterium glutamicum*	[42]
6-Aminocaproic acid	0.16	—	Glucose	*Escherichia coli*	[44]
Glutaconic acid	0.35	—	Glucose	*Escherichia coli*	[4]
Succinic acid	31.9	0.78	Glucose	*Escherichia coli*	[46]
	86.6	0.92	Glucose	*Escherichia coli*	[47]
	40	—	Glucose	*Escherichia coli*	[49]
	58.3	0.62	Glucose	*Escherichia coli*	[50]
	3.6	0.07	Glucose	*Saccharomyces cerevisiae*	[51]
	13	0.14	Glucose	*Saccharomyces cerevisiae*	[52]
1,4-Butanediol	18	—	Glucose	*Escherichia coli*	[53]
	29	—	Glucose	*Escherichia coli*	[54]
	99	0.35	Glucose	*Escherichia coli*	[54]

(Continued)

Table 2.1.   *(Continued)*

Product	Titer (g/L)	Yield (g/g)	Carbon source	Host	References
Fumaric acid	22.8	0.35	Glucose and crude glycerol	*Rhizopus oryzae*	[58]
	24	0.78	Glucose	*Rhizopus oryzae*	[59]
	41.5	0.44	Glycerol	*Escherichia coli*	[61]
	28.2	0.39	Glucose	*Escherichia coli*	[62]
	5.7	0.13	Glucose	*Saccharomyces cerevisiae*	[63]
	33.1	—	Glucose	*Saccharomyces cerevisiae*	[64]
	15.8	—	Glucose	*Candida glabrata*	[65]
β-Alanine	32.3	0.14	Glucose	*Escherichia coli*	[68]
3-Hydroxypropionic acid	31.1	0.42	Glucose	*Escherichia coli*	[69]
Malonic acid	3.6	0.04	Glucose	*Escherichia coli*	[69]
Malic acid	113	0.95	Glucose	*Aspergillus flavus*	[72]
	9.3	0.56	Glucose	*Escherichia coli*	[73]
	34	1.06	Glucose	*Escherichia coli*	[74]
	59	0.31	Glucose	*Saccharomyces cerevisiae*	[75]
	142.2	0.13	Glucose	*Aureobasidium pullulans*	[76]

citric acid.[8] After that, a lot of efforts have been made aiming to increase the titer of citric acid in *A. niger.*

Overexpression of rate-limiting enzymes involved in the producing pathways is the commonly used method to increase the production of target products. However, it has little effect on citric acid production caused by the tight regulation of central carbon metabolism. Ruijter reported that overexpression of phosphofructokinase and pyruvate kinase to increase the flux of glycolytic pathway in *A. niger* produced 55 g/L of citric acid (Table 2.1), which was the same with wild-type strain as well.[9] Similarly, overexpression of citrate synthase also failed to increase the titer of citric acid.[10]

Basically, the production of citric acid of *A. niger* relies on the high glycolytic flux. An effective strategy to increase the flux of glycolytic is to accumulate ADP and NAD^+.[11,12] Wang blocked the ATP synthesis by adding antimycin A or DNP into the medium, and the citric acid titer rose to 151 g/L and 135 g/L (Table 2.1), respectively. The production increased by 20% and 7%, respectively.[13]

Elimination of inhibitor effect and removal of by-product are also used to improve citric acid titer. Arisan found that *A. niger* initiated citric acid production earlier by deletion of trehalose-6-phosphate synthase A to decrease the concentration of trehalose-6-phosphate, an inhibitor of hexokinase.[14] Ruijter deleted glucose oxidase and oxaloacetate acetylhydrolase to eliminate oxalic acid production, which successfully increased the citric acid titer.[15]

Alternatively, the yeast strain *Yarrowia lipolytica* has also been engineered for overproduction of citric acid. When stearin and glucose were used as the cosubstrates, *Y. lipolytica* produced 43 g/L of citric acid[16] (Table 2.1). The recombinant *Y. lipolytica*, harboring invertase from *Saccharomyces cerevisiae* and multiple copies of its own isocitrate lyase, produced 140 g/L of citric acid (Table 2.1) using sucrose as the sole carbon source.[17] Liu deleted the ATP-citrate lyase and overexpressed isocitrate lyase, leading to 84 g/L of citric acid produced from 10% inulin[18] (Table 2.1).

2.6.2. *Chemical Derived from Citric Acid*

Itaconic acid is derived from citric acid catalyzed by aconitase and *cis*-aconitate decarboxylase (Figure 2.3). As a sustainable industrial building block, it is widely used in coatings, plastics, detergents, and rubber industries.[19] The market size of itaconic acid is estimated at 410,000 tons per year in 2020.[20]

Industrially, *Aspergillus terreus* is the best producer for overproducing itaconic acid. Up to now, mutagenesis and process condition optimization are still the dominant methods to increase the itaconic acid production in *A. terreus*, which achieved titers of 80–150 g/L.[21,22] However, those processes usually need to be further optimized to reduce the cost of itaconic acid production, which can be accomplished by achieving higher titer and yields, reducing the fermentation time, utilizing cheaper and sustainable substrates. This can be addressed through metabolic engineering.

Actually, great progress has been made in recent years. Huang demonstrated that overexpression of *cis*-aconitate decarboxylase and the major facilitator superfamily transporter *mfsA* in *A. terreus* increased the titer of itaconic acid by 9.4% and 5.1%, respectively.[23] Using glucose as the carbon source leads to high production cost, while using lignocellulosic raw material as the carbon source is a promising alternative for cutting the cost. Kocabas investigated utilization of agricultural residues as the carbon source in *A. terreus* to enable production of 18 g/L itaconic acid by applying two-step fermentation[24] (Table 2.1).

Compared with *A. terreus*, *Escherichia coli* is more suitable for producing itaconic acid because of its higher growth rate and ability to manipulate easily. Okamoto reported that titer of up to 4.3 g/L (Table 2.1) was achieved by heterologous overexpressing of *cis*-aconitate decarboxylase and native aconitase and inactivating isocitrate dehydrogenase gene in *E. coli*.[25] The highest itaconic acid titer reported so far is 32 g/L in *E. coli* in fed-batch fermentation (Table 2.1). This was achieved by applying a model-based approach to select target genes. Deletion of those genes has successfully redirected the carbon flux into itaconic acid production.[26]

2.7. α-Ketoglutaric Acid and Its Derived Compounds

2.7.1. α-Ketoglutaric Acid

α-Ketoglutaric acid is a key intermediate in the citric acid cycle and amino acids as well as protein metabolism, which is synthesized by isocitrate dehydrogenase-catalyzed oxidative decarboxylation of isocitrate. It has a wide range of applications in food, agrochemical, and pharmaceutical industries. For example, it is a building block for the industrial synthesis of heterocycles.[27] As a popular nutrition enhancer, α-ketoglutaric acid is used to improve athletic performance.[28] It also can improve gut morphology and function and aids wound healing.[29]

A large number of microorganisms were selected to produce α-ketoglutaric acid. Among them, the yeast *Yarrowia lipolytica* has been placed as a superior host due to its high product yield, broad-spectrum substrate activity, and simple process control.[30] In the early times, *n*-paraffin was the commonly used carbon source for overproducing α-ketoglutaric acid in *Y. lipolytica*. As shown in Table 2.1, the highest titer was 195 g/L by using a mixture of *n*-paraffin (C_{12}–C_{18}) as the substrate.[31] After that, only a few articles reported using *n*-paraffin as the carbon source because of its high price and limited supply. Recent studies reported utilization of ethanol, glycerol, and vegetable oils as the carbon source to produce α-ketoglutaric acid in *Y. lipolytica*.[32] However, those engineered strains have not been applied in industrial manufacturing caused by lots of by-products and unwanted carboxylates such as pyruvate, fumarate, malate, and succinate that accumulate in the medium. Several metabolic engineering strategies have been used to address this issue.

The production of α-ketoglutaric acid has been increased by overexpression of key enzymes from the carbon assimilation pathway and elimination of by-products. Overexpression of pyruvate carboxylation pathway by overexpressing pyruvate carboxylase successfully redirected the carbon flux from pyruvate to α-ketoglutaric acid. The final titer reached 62.5 g/L and the by-product pyruvate decreased from 35.2 to 13.5 g/L.[30] Overexpression of $NADP^+$-dependent isocitrate dehydrogenase and pyruvate carboxylase enhanced the flux of citric acid cycle and simultaneously decreased the production of pyruvate, producing 186 g/L of α-ketoglutaric acid in the bioreactor.[27] The effect of fumarase and pyruvate carboxylase on α-ketoglutaric acid production has also been investigated. The highest titer reached 132.7 g/L by overexpression of both of those enzymes[33] (Table 2.1).

Another effective strategy to improve the titer of α-ketoglutaric acid is to supply or regenerate cofactors. Zhou reported that heterologous overexpression of acetyl-CoA synthetase from *S. cerevisiae* and ATP-citrate lyase gene from *Mus musculus* increased the intracellular level of acetyl-CoA and led to an enhanced α-ketoglutaric acid production at a titer of 56.5 g/L in a 3-L fenmentor.[34] Cofactor thiamine regulates the enzymatic activity of pyruvate dehydrogenase complex and ketoglutarate dehydrogenase complex. The former plays an important role in supplying the key precursor acetyl-CoA. Thiamine utilization between those two enzymes was distributed by overexpression of pyruvate dehydrogenase complex E1 component α-subunit, which is a TPP-binding subunit. The engineered strain accumulated less pyruvate with an enhanced α-ketoglutaric acid production at a titer of 43.3 g/L[35] (Table 2.1).

2.7.2. Chemicals Derived from α-Ketoglutaric Acid

As shown in Figure 2.4(a), α-ketoglutaric acid as a building block can be used to synthesize a wide range of value-added chemicals through various biochemical reactions such as transamination, decarboxylation, reduction, and chain elongation.[36]

L-Glutamate is synthesized by the reduction of or transamination of α-ketoglutaric acid catalyzed by glutamate dehydrogenase or transaminase, respectively. It is an important amino acid used to produce the flavor enhancer monosodium glutamate. The current market size of L-glutamate is 2.3 million tons per year, and it is commonly produced by *Corynebacterium glutamicum*.[20] Increasing the intracellular level of precursor α-ketoglutaric acid by overexpression of upstream rate-limiting enzymes and deletion of competing pathways are the main strategies used to improve L-glutamate production. Overexpression of pyruvate carboxylase and deletion of pyruvate kinase and α-ketoglutaric acid dehydrogenase or inhibition of α-ketoglutaric acid dehydrogenase by antisense RNA or enzyme inhibitor successfully increased L-glutamate titer.[37-40] Another effective strategy is to enhance the citric acid cycle by heterologous overexpression of polyhydroxybutyrate (PHB) synthase from *Ralstonia eutropha* or hemoglobin gene *vgb* from *Vitreoscilla* in *C. glutamicum*. The L-glutamate production increased by 23% and 22%, respectively.[3,41] The improvement of L-glutamate titer also can be achieved by introducing the phosphoketolase pathway into *C. glutamicum* to reduce carbon dioxide emission level.[42]

(a)

HOOC — 2-Oxopimelate — COOH ← Chain elongation 2 — HOOC — α-Ketoglutaric acid — COOH → Reduction 5 → HOOC — (R)-2-Hydroxyglutarate — COOH

↓ 3 | Transamination 1 ↓ NH$_2$ | ↓ 6

6-Oxohexanoate | Glutamic acid | CoAS — (R)-2-Hydroxyglutaryl-CoA — COOH

↓ 4 | | ↓ 7

Adipic acid | | CoAS — Glutaconyl-CoA — COOH

| | ↓ 8

| | HOOC — Glutaconic acid — COOH

(b)

Succinic acid   HOOC — α-Ketoglutaric acid — COOH

↓ 9 | ↘ 10

Succinyl CoA (SCoA)

↓ 11 ↘ | ↙

Succinyl semialdehyde

↓ 12

4-Hydroxybutyrate

↓ 13

4-Hydroxybutyryl CoA (CoAS)

↓ 14

4-Hydroxybutyraldehyde

↓ 15

1,4-Butanedio

(c)

Oxaloacetic acid   Fumaric acid

↘ 16 | 17 ↙

Aspartic acid (NH$_2$)

↓ 18

β-Alanine (H$_2$N)

↓ 19

Malonic semialdehyde

↙ 20 | 21 ↘

Malonic acid   3-Hydroxypropionic acid

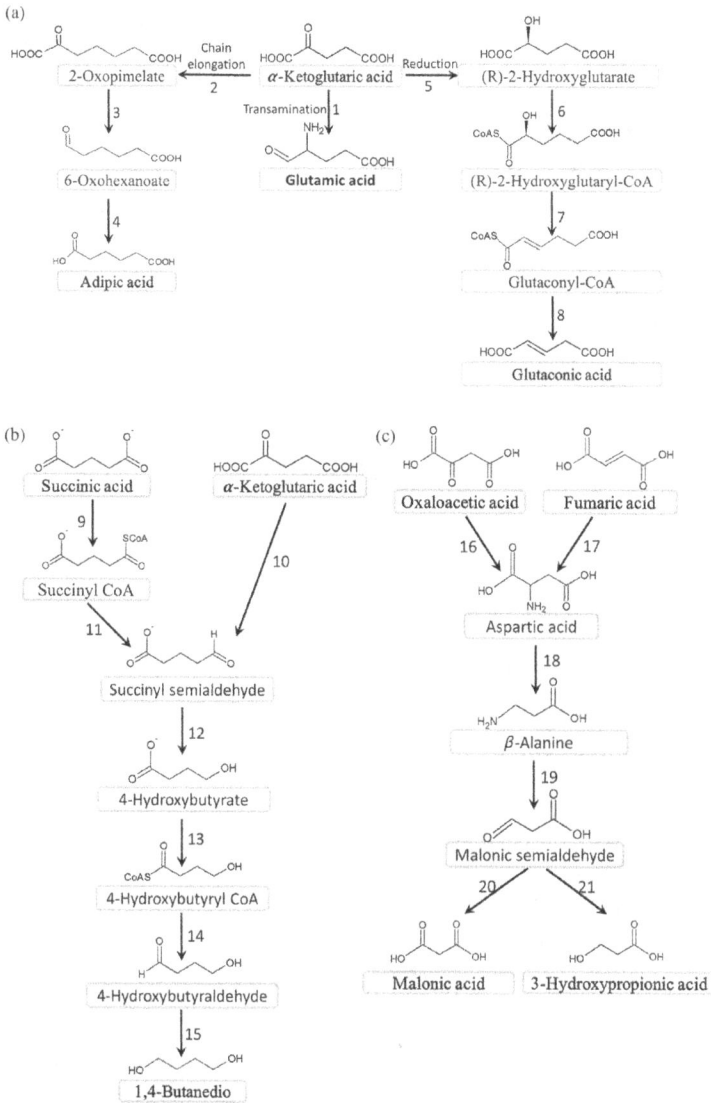

Figure 2.4. Metabolic pathway for the production of high-value derivatives from citric acid cycle. (a): The metabolic pathways from α-ketoglutaric acid to adipic acid, glutamic acid, and glutaconic acid, respectively. 1: Glutamate dehydrogenase or transaminase; 2: Keto acid elongation enzymes include homocitrate synthase, homoaconitase a subunit, homoaconitase b subunit and threo-isohomocitrate dehydrogenase; 3: Keto acid decarboxylase; 4: Aldehyde dehydrogenase; 5: 2-Hydroxyglutarate dehydrogenase; 6 and 8: Glutaconate CoA transferase; 7: 2-Hydroxyglutaryl-CoA dehydratase.

Adipic acid is an important building block used for the synthesis of nylon and polyurethane foam and registers an annual demand of 2.8 metric tons.[5] Recently, a patent proposed a biosynthetic pathway to produce adipic acid. α-Ketoglutaric acid was elongated to obtain α-ketopemilic acid by introducing keto acid elongation enzymes AksADEF involved in coenzyme B biosynthetic pathway in *Methanogenic archaea*. Adipic acid was produced by carboxylation and oxidation of α-ketopemilic acid.[43] In another report, α-ketopemilic acid was transaminated and further decarboxylated to 6-aminocaproic acid, a precursor of synthetic polymer nylon-6[44] [Figure 2.4(a)].

Glutaconic acid can be reduced to glutaric acid, both of which are commonly used as the monomers to synthesize polyesters and polyamides.[4] Glutaconic acid was produced from α-ketoglutaric acid by assembling genes encoding 2-hydroxyglutarate dehydrogenase, glutaconate CoA-transferase, and 2-hydroxyglutaryl-CoA dehydratase into *E. coli* [Figure 2.4(a)]. The engineered host strain produced 350 mg/L glutaconic acid when cultured on a complex medium.[4]

2.8. Succinic Acid and Its Derived Compounds

2.8.1. *Succinic Acid*

Succinic acid is an intermediate in citric acid cycle produced from α-ketoglutaric acid catalyzed by α-ketoglutaric acid dehydrogenase and SUCLG. It has received great interest in recent years as a chemical building block for a range of applications such as foods, plastics, detergents, solvents, resins, and polymers.[5] It also can be used as the precursor to

---

Figure 2.4. (*Continued*) (b): The metabolic pathway from succinic acid and α-ketoglutaric acid to 1,4-butanediol.[53] 9: SUCLG; 10: 2-Oxoglutarate decarboxylase; 11: CoA-dependent succinate semialdehyde dehydrogenase. 12: 4-Hydroxybutyrate dehydrogenase; 13: 4-Hydroxybutyryl-CoA transferase; 14: 4-Hydroxybutyryl-CoA reductase; 15: Alcohol dehydrogenase.

(c): The metabolic pathway from oxaloacetic acid and fumaric acid to malonic acid and 3-hydroxypropionic acid.[69] 16: Aspartate aminotransferase; 17: Aspartase; 18: Aspartate-α-decarboxylase; 19: β-Alanine pyruvate transaminase; 20: Succinic semialdehyde dehydrogenase; 21: Malonic semialdehyde reductase.

synthesize many bulk chemicals including gamma-butyrolactone, 1,4-butanediol, *n*-methylpyrrolidone, and tetrahydrofuran.[1] The global annual production of succinic acid has been estimated at 30,000–50,000 tons.[45]

Several bacteria such as *Anaerobiospirillum succiniciproducens* and *Actinobacillus succinogenes* naturally secrete a large quantity of succinic acid, respectively, producing 50 and 90 g/L during anaerobic fermentation.[1] However, the unstable productivity and strict anaerobic culture conditions make those strains unfit for industrial application.

E. coli has been metabolically engineered for high-yield succinic acid production through the reductive citric acid cycle under anaerobic conditions. The elimination of by-products such as ethanol, lactate, and acetate by gene knockout is the major strategy used to improve the titer. However, those deletions lead to poor growth of *E. coli* on glucose under anaerobic condition due to insufficient cellular energy and electron acceptors. This issue can be addressed by overexpression of malate dehydrogenase to supplement of NAD^+, and the engineered strain produced 31.9 g/L of succinic acid with a yield of 1.19 mol/mol glucose[46] (Table 2.1). Another approach was taken to restore cell growth through metabolic evolution and selection. The obtained mutant with deletion of *focA* and *pflB* produced 733 mM (86.6 g/L) succinic acid with a high yield of 1.41 mol/mol glucose.[47] Additional deletion of *mgsA, poxB, tdcD,* and *tdcE* reduced the acetate yield by 50% and increased the succinic acid yield by 10%.[48]

Enhancement of glyoxylate shunt pathway has also been used to improve succinic acid titer. Under anaerobic condition, 40 g/L of succinic acid was produced by activating glyoxylate shunt pathway via deletion of the glyoxylate shunt transcriptional repressor *iclR* in an engineered *E. coli* strain, which lacks *adhE, ldhA,* and *pta-ackA*. Further engineering including heterologous expression of *pyc* from *Lactococcus lactis* and *fdh1* from *Candida boidinii* to improve the intracellular NADH level increased the yield of succinic acid by 6%.[49] Under aerobic condition, glyoxylate shunt pathway was constructed by inactivation of *ptsG, sdhAB,* and *iclR,* together with blocking of by-product formation (*poxB* and *pat-ackA*) and overexpression of *ppc* from *Sorghum vulgare* resulted in 58.3 g/L succinic acid at a yield of 0.94 mol/mol glucose[50] (Table 2.1).

S. *cerevisiae* is a promising host for overproducing succinic acid owing to low pH fermentation tolerance. Redirecting carbon flux into glyoxylate shunt and reducing succinic acid consumption by deletion of *SDH1*, *SDH2*, *IDH1*, and *IDP1* to inactivate succinic acid and isocitric acid dehydrogenase produced 3.62 g/L of succinic acid at a yield of 0.11 mol/mol glucose[51] (Table 2.1). In a pyruvate decarboxylase-deficient S. *cerevisiae*, fumarate hydratase (*fum1*) was replaced by E. *coli* fumarate hydratase *fumC* to block the malate synthesis from fumarate. Glycerol-3-phosphate dehydrogenase (*GPD1*) also deleted to maintain the redox balance. Besides, pyruvate carboxylase (*PYC*), malate dehydrogenase (*MDH*), and fumarate reductase (*FRD*) were overexpressed. This engineered strain achieved succinic acid titer of 13 g/L with supplement of urea, $CaCO_3$, and biotin.[52] Although succinic acid titer and yield are lower in S. *cerevisiae* compared with other producers, it demonstrated that metabolic engineering approach provides a promising proof-of-concept.

2.8.2. *Chemical Derived from Succinic Acid*

1,4-Butanediol is used industrially in the manufacture of plastics, elastic fibers, and polyesters at a scale of over 2.5 million tons per year.[5] Yim first reported the biosynthesis of 1,4-butanediol from citric acid cycle intermediates, α-ketoglutaric acid and succinic acid in E. *coli*[53] [Figure 2.4(b)]. Both of them can be converted to the same precursor succinyl semialdehyde catalyzed by exogenous α-ketoglutaric acid decarboxylase and succinate semialdehyde dehydrogenase, respectively. The last three steps are responsible for the conversion of succinyl semialdehyde to 1,4-butanediol by heterologous overexpression of 4-hydroxybutyrate dehydrogenase, 4-hydroxybutyryl-CoA transferase, and alcohol dehydrogenase. Introducing those enzymes into wild-type E. *coli*, 1.2 g/L of 1,4-butanediol, along with other by-products, was obtained. Metabolic engineering optimizations aimed at eliminating by-products and enhancing reducing power were conducted by inactivation of *ldhA*, *pflB*, *adhE*, and *mdhA,* leading to 18 g/L of 1,4-butanediol produced from glucose in bioreactor.[53] However, it still accumulated high levels of pyruvate and acetate. [13]C labeling experiments and transcriptomics study suggested that the carbon flux could go back into citric acid cycle from the 1,4-butanediol pathway in the reaction of succinate semialdehyde dehydrogenases encoded by *sad* and *gabD*. Deletion of those two genes led to 1,4-butanediol titer to 29 g/L.[54] Industrially relevant production of 1,4-butanediol has been achieved by further improving the performance of the engineered strain, including efficient enzyme discovery and fermentation process optimization. The highest titer reached to 99 g/L with a yield of 0.35 g/g glucose[54] (Table 2.1).

2.9. Fumaric Acid and Its Derived Compounds

2.9.1. *Fumaric Acid*

Fumaric acid was produced by the oxidation of succinic acid catalyzed by SDH in the citric acid cycle. It is an unsaturated dicarboxylic acid, and therefore can be used as a starting material for esterification and polymerization reactions to synthesize plastics, esters, resins, and polymers.[55] It is also used as additives in food and beverage because of its safety and acidic taste.[56] Another potential application of fumaric acid is as an antibacterial factor and physiologically active agent in the feed industry.[56] Additionally, as an important platform chemical, fumaric acid has been used to prepare edible acids including L-malic acid and L-aspartic acid.[57]

Biosynthesis of fumaric acid mainly uses the reverse citric acid cycle, containing three key enzymes. Pyruvate carboxylase is responsible for carboxylation of pyruvate to oxaloacetate with the participation of ATP and CO_2. Then the malate dehydrogenase converts oxaloacetate to malic acid. Fumaric acid was produced from malate by the enzyme fumarase.

Biological production of fumaric acid was traditionally carried out using the best producer *Rhizopus* species. A titer value of 22.81 g/L for fumaric acid can be obtained by coutilization of glucose and crude glycerol as the carbon source in *R. arrhizus*[58] (Table 2.1). The strategy of overexpressing key enzymes has been used to increase fumaric acid production. On heterologous overexpression of phosphoenolpyruvate carboxylase (PEPC) from *E. coli* to redirect carbon flux into oxaloacetate led to 24 g/L of fumaric acid with a yield of 0.78 g/g glucose in *R. oryzae*. The titer improved by 26% compared with the wild-type strain.[59] In another report, overexpression of fumarase (*fumR*) has been expected to reduce the by-product malic acid. However, the engineered strain produced more malic acid rather than fumaric acid, which indicated that fumarase is not responsible for increasing fumaric acid production in *R. oryzae*.[60] Actually, there are some challenges for employing *R. oryzae* to industrial manufacture of fumaric acid due to their morphological characteristics of forming cell aggregates. There is a need to develop other hosts for fumaric acid production.

The well-known, industrial workhorse *E. coli* has been metabolically engineered for fumaric acid production. Deletion of fumarases and over-expression of PEP carboxylase (*ppc*) or the glyoxylate shunt operon (*aceBA*) enabled *E. coli* to accumulate 41.5 g/L of fumaric acid from glycerol in fed-batch culture[61] (Table 2.1). A recent study reported that engineered *E. coli* produced 28.2 g/L of fumaric acid by combining inactivation of fumarases (*fumA, fumB,* and *fumC*) and *iclR* to redirect carbon flux into glyoxylate shunt, overexpression of *ppc* to enhance reverse citric acid cycle flux, and inactivation of *arcA* and *ptsG* to increase the oxidative citric acid cycle flux[62] (Table 2.1).

The high acid-tolerance yeast strain has also been engineered to produce fumaric acid. By simultaneously introducing reverse and oxidative citric acid cycle into *S. cerevisiae* via deletion of *THI2* and *FUM1* and overexpression of exogenous *PYC, MDH* and *FUM1* form *R. oryzae*, 5.65 g/L of fumaric acid was obtained.[63] Recently, a relatively high fumaric acid titer of 33.13 g/L was achieved through modular pathway engineering in *S. cerevisiae*.[64] A total of 10 essential genes involved in fumaric acid biosynthetic pathway were arranged into three modules: reduction module, oxidation module, and by-product module. A systematic optimization of those modules enabled an efficient microbial production of fumaric acid.[64] The oxidative pathway for fumaric acid production was constructed in *Candida glabrata* by overexpression of KGD complex, SUCLG, and SDH. Fine-tuning their expression level and overexpression of two dicarboxylic acids transporters resulted in fumaric acid titer of 15.76 g/L.[65]

2.9.2. Chemicals Derived from Fumaric Acid

β-Alanine, also called 3-aminopropionic acid, is a naturally occurring β-amino acid. It is a potential intermediate for the production of acrylamide and acrylonitrile[66] and a precursor for the synthesis of poly-β-alanine, which is applied in cosmetics, water purification, and construction.[67] Recently, a fumaric acid overproducing *E. coli* strain has been engineered to secrete a high level of β-alanine at a titer of 32.3 g/L [Table 2.1 and Figure 2.4(c)]. This was achieved by overexpressing L-aspartate-α-decarboxylase (*panD*) from *C. glutamicum* and reinforcing the expression level of aspartase and PEPC.[68] This pathway has been further engineered to produce two high-value products, 3-hydroxypropionic acid and malonic acid[69] [Figure 2.4(c)]. 3-Hydroxypropionic acid was selected as a top building block compound by the U.S. Department of Energy. Many three-carbon intermediates are derived from 3-hydroxypropionic acid such as acrylic acid, 1,3-propanediol, acrylamide, 3-hydroxypropylamine, β-propiolactone, and malonic acid.[20] Malonic acid is another top chemical, which has been used in a wide range of industries including polymer crosslinking, resins, electronics, flavors and fragrances, solvents, and pharmaceuticals.[69] By heterologous expression of β-alanine pyruvate transamimase from *Pseudomonas aeruginosa,* the β-alanine pathway has been extended to produce malonic semialdehyde, which could be reduced to 3-hydroxypropionic acid or oxidized to malonic acid. The final engineered strains produced 31.1 g/L of 3-hydroxypropionic acid or 3.6 g/L of malonic acid by introducing *E. coli* malonic semialdehyde reductase (*ydfF*) or semialdehyde dehydrogenase (*yneI*).[69]

2.10. Malic Acid and Its Derived Compounds

Malic acid is a dicarboxylic acid, an intermediate derived from fumaric acid catalyzed by fumarase in the citric acid cycle. It can be used in the food industry as an acidulant and flavor enhancer. Malic acid is also used in metal cleaning and finishing, pharmaceuticals, agriculture, and polymers.[70] Demand for this high-value compound is currently around 60,000 tons per year with a growth rate of 4%.[71]

Aspergillus flavus is a well-known malic acid producer. The higher titer reported in *A. flavus* is 113 g/L with a yield of 0.95 g/g glucose and a productivity of 0.59 g/L/h[72] (Table 2.1). However, *A. flavus* is not suitable for industrial manufacturing due to oxygen transfer and aflatoxin production.

Biosynthesis of malic acid has been achieved in *E coli*. A titer of 9.25 g/L of malic acid was obtained by introducing PEP carboxykinase (*pckA*) from *Mannheimia succiniciproducens*[73] (Table 2.1). In another report, a succinic acid overproducing *E. coli* strain was further engineered to secrete a high level of malic acid.[74] The carbon flux was redirected into malic acid by inactivation of fumarase isoenzymes (*fumB* and *fumAC*) and fumarase reductase (*frdBC*), combined with elimination of by-products (ethanol, acetate, and lactate). The final strain produced 34 g/L of malic acid with a yield of 1.06 g/g glucose.[74]

Other hosts have also been engineered to accumulate malic acid such as *S. cerevisiae* and *Aureobasidium pullulans*.[75,76] In the case of *S. cerevisiae*, overexpression of endogenous *PYC* and *MDH*, combined with introduction of a malic acid transporter (Sp*MAE1*) from *Schizosaccharomyces pombe*, in a pyruvate decarboxylase-deficient *S. cerevisiae*, resulted in 59 g/L of malic acid with a yield of 0.31 g/g glucose.[75] Recently, a polymalic acid overproducing strain *A. pullulans* has been developed to produce malic acid by acid hydrolysis. This engineered strain produced 142.2 g/L of malic acid with a productivity of 0.74 g/L/h in fed-batch fermentation, and has shown potential for industrial application[76] (Table 2.1).

2.11. Oxaloacetic Acid and Its Derived Compounds

Oxaloacetic acid is formed from oxidation of malic acid, catalyzed by malate dehydrogenase in citric acid cycle. It can be used as a precursor for the synthesis of amino acids such as aspartate, alanine, asparagine, methionine, lysine, and threonine. Production of oxaloacetate in *E. coli* has been achieved by heterologous expression of codon-optimized PEPC from *Dunaliella salina* or *Photobacterium profundum* SS9.[77,78] Recently, an engineered *S. cerevisiae* strain produced 13.7 g/L of 3-hydroxypropionic acid by using oxaloacetate as the precursor. In their work, oxaloacetate was converted into aspartic acid and then into the *β*-alanine catalyzed by aspartate aminotransferase and aspartate decarboxylase, respectively. *β*-Alanine was further converted into 3-hydroxypropionic acid by introducing *β*-alanine-pyruvate aminotransferase from *Bacillus cereus* and 3-hydroxypropanoate dehydrogenase (*ydfG*) from *E. coli*.[79]

2.12. Conclusion and Perspective

Citric acid cycle intermediates and their derivatives are building block chemicals that can be used in a wide range of applications. Due to their huge market size and value, a lot of efforts have been made to directed at highly efficient microorganisms that can produce a significant amount of those intermediates in recent years. Many traditional metabolic engineering strategies have succeeded in developing microorganisms capable of over-producing those products, including (1) metabolic evolution to obtain desired phenotypes, (2) substrate utilization engineering to achieve inexpensive carbon source utilization, (3) precursor enrichment and by-product elimination to increase the flux toward the products, (4) transporter engineering to facilitate product secretion, and (5) cofactor optimization and inhibitor elimination to improve enzyme activity. However, relatively few such engineered strains are ever scaled up to achieve industrial manu-facturing, due to the lower titer, yield, and productivity compared with chemical process. In this regard, various hurdles need to be overcome.

First, the strains developed in laboratories do not perform well in industrial fermentation due to their genetic instability. Chromosomal inte-gration is a preferable strategy to overcome this drawback, which can be achieved by homologous recombination, site-specific recombination, transposon-mediated gene transposition, and CRISPR-based genome editing.[81,82]

Second, the overall cost of bioprocessing is still staying at a relatively high level. The use of antibiotics and isopropyl β-D-1-thiogalactopyrano-side (IPTG) should be avoided in the fermentation process. This can be addressed by using chromosomal integration to replace plasmid-based expression system and using constitutive promoters or synthetic biology based auto-inducer circuits rather than the chemical inducer IPTG.[83,84] Additionally, the host strains should be engineered to use more sustaina-ble and less expensive substrates, such as food and agriculture wastes, lignin, and lignocellulosic hydrolysates. Although recent progress has been achieved in using carbon dioxide, protein waste, methanol, and methane as the carbon source to produce high-value compounds,[85–89] it is still necessary to advance further to make those feedstocks more economi-cally competitive with glucose or sucrose.

Third, some industrial microorganisms cannot tolerate higher product concentrations. It is difficult to enhance product tolerance because our limited understanding of the molecular mechanisms involved. Transporter engineering combined with adaptive evolution is a promising strategy to solve this problem. This approach has successfully improved fatty acids and butanol tolerance to *E. coli* and 3-hydroxypropionic acid tolerance to *S. cerevisiae*.[90,91]

Fourth, it is complicated to optimize carbon flux due to the trade-off between cell growth and target product formation, as well as the complexity of cofactor and inhibitor regulation. Metabolic network modeling and simulation at genome scale provides a promising method to reveal the key elements of a metabolic network, thereby improving the performance of the host strain.[92,93] Besides, whole-cell computational modeling is able to capture cellular metabolism as well as regulatory and signaling circuits, which can be used to facilitate metabolic flux optimization.[94]

Even though several challenges still need to be overcome, it is expected that microbial production of citric acid cycle derived products from biomass will gradually replace the petroleum-based chemical process in the future. Fortunately, with rapid progress in synthetic biology and metabolic engineering, many tools such as expression system controlling, genome editing, and high-throughput screening have been developed to help us construct microorganism cell factories. We believe that with aids of those tools and increasing knowledge on metabolic systems, a more sustainable and economic society will be on the horizon.

References

1. Chen, Y., & Nielsen, J. Biobased organic acids production by metabolically engineered microorganisms. *Curr. Opin. Biotechnol.* **17**, 165–172 (2016).
2. Yin, X. *et al.* Metabolic engineering in the biotechnological production of organic acids in the tricarboxylic acid cycle of microorganisms: Advances and prospects. *Biotechnol. Adv.* **33**, 830–841 (2015).
3. Liu, Q. *et al.* Microbial production of L-glutamate and L-glutamine by recombinant *Corynebacterium glutamicum* harboring *Vitreoscilla* hemoglobin gene *vgb*. *Appl. Microbiol. Biotechnol.* **77**, 1297–1304 (2008).

4. Djurdjevic, I., Zelder, O., & Buckel, W. Production of glutaconic acid in a recombinant *Escherichia coli* strain. *Appl. Environ. Microbiol.* **77**, 320–322 (2011).

5. Sun, X. *et al.* Synthesis of chemicals by metabolic engineering of microbes. *Chem. Soc. Rev.* **44**, 3760–3785 (2015).

6. Kobayashi, K., Hattori, T., Hayashi, R., & Kirimura, K. Overexpression of the NADP$^+$-specific isocitrate dehydrogenase gene (*icdA*) in citric acid-producing *Aspergillus niger* WU-2223L. *Biosci. Biotechnol. Biochem.* **78**, 1246–1253 (2014).

7. Angumeenal, A.R., & Venkappayya, D. An overview of citric acid production. *LWT — Food Sci. Technol.* **50**, 367–370 (2013).

8. Zahorsky, B. U.S. Patent No. 1065358 (1913).

9. Ruijter, G.J.G., Panneman, H., & Visser, J. Overexpression of phosphofructokinase and pyruvate kinase in citric acid-producing *Aspergillus niger*. *BBA — Gen. Subjects* **1334**, 317–326 (1997).

10. Ruijter, G.J.G., Panneman, H., Xu, D.-B., & Visser, J. Properties of *Aspergillus niger* citrate synthase and effects of *citA* overexpression on citric acid production. *FEMS Microbiol. Lett.* **184**, 35–40 (2000).

11. Thomas, S., & Fell, D.A. A control analysis exploration of the role of ATP utilisation in glycolytic-flux control and glycolytic-metabolite-concentration regulation. *Eur. J. Biochem.* **258**, 956–967 (1998).

12. Neves, A.R. *et al.* Is the glycolytic flux in *Lactococcus lactis* primarily controlled by the redox charge? Kinetic of NAD$^+$ and NADH pools determined *in vivo* by ^{13}C NMR. *J. Biol. Chem.* **277**, 28088–28098 (2002).

13. Wang, L. *et al.* Inhibition of oxidative phosphorylation for enhancing citric acid production by *Aspergillus niger*. *Microb. Cell Fact.* **14**(1), 7, doi:10.1186/s12934-12015-10190-z (2015).

14. Arisan-Atac, I., Wolschek, M.F., & Kubicek, C.P. Trehalose-6-phosphate synthase A affects citrate accumulation by *Aspergih niger* under conditions of high glycolytic flux. *FEMS Microbiol. Lett.* **140**, 77–83 (1996).

15. Ruijter, G.J., van de Vondervoort, P.J., & Visser, J. Oxalic acid production by *Aspergillus niger*: An oxalate-non-producing mutant produces citric acid at pH 5 and in the presence of manganese. *Microbiology* **145**, 2569–2576 (1999).

16. Papanikolaou, S. *et al.* Influence of glucose and saturated free-fatty acid mixtures on citric acid and lipid production by *Yarrowia lipolytica*. *Curr. Microbiol.* **52**, 134–142 (2006).

17. Förster, A., Aurich, A., Mauersberger, S., & Barth, G. Citric acid production from sucrose using a recombinant strain of the yeast *Yarrowia lipolytica*. *Appl. Microbiol. Biotechnol.* **75**, 1409–1417 (2007).

18. Liu, X.-Y., Chi, Z., Liu, G.-L., Madzak, C., & Chi, Z.-M. Both decrease in *ACL1* gene expression and increase in *ICL1* gene expression in marine-derived yeast *Yarrowia lipolytica* expressing *INU1* gene enhance citric acid production from inulin. *Mar. Biotechnol.* **15**, 26–36 (2013).

19. Hajian, H., & Yusoff, W.M.W. Itaconic acid production by microorganisms: A review. *Curr. Res. J. Biol. Sci.* **7**, 37–42 (2015).

20. Choi, S., Song, C.W., Shin, J.H., & Lee, S.Y. Biorefineries for the production of top building block chemicals and their derivatives. *Metab. Eng.* **28**, 223–239 (2015).

21. Kuenz, A., Gallenmüller, Y., Willke, T., & Vorlop, K.-D. Microbial production of itaconic acid: Developing a stable platform for high product concentrations. *Appl. Microbiol. Biotechnol.* **96**, 1209–1216 (2012).

22. Hevekerl, A., Kuenz, A., & Vorlop, K.-D. Influence of the pH on the itaconic acid production with *Aspergillus terreus*. *Appl. Microbiol. Biotechnol.* **98**, 10005–10012 (2014).

23. Huang, X., Lu, X., Li, Y., Li, X., & Li, J.-J. Improving itaconic acid production through genetic engineering of an industrial *Aspergillus terreus* strain. *Microb. Cell Fact.* **13**, doi:10.1186/s12934-12014-10119-y (2014).

24. Kocabas, A., Ogel, Z.B., & Bakir, U. Xylanase and itaconic acid production by *Aspergillus terreus* NRRL 1960 within a biorefinery concept. *Ann. Microbiol.* **64**, 75–84 (2014).

25. Okamoto, S. *et al.* Production of itaconic acid using metabolically engineered *Escherichia coli.* *J. Gen. Appl. Microbiol.* **60**, 191–197 (2014).

26. Harder, B.-J., Bettenbrock, K., & Klamt, S. Model-based metabolic engineering enables high yield itaconic acid production by *Escherichia coli.* *Metab. Eng.* **38**, 29–37 (2016).

27. Yovkova, V., Otto, C., Aurich, A., Mauersberger, S., & Barth, G. Engineering the α-ketoglutarate overproduction from raw glycerol by overexpression of the genes encoding $NADP^+$-dependent isocitrate dehydrogenase and pyruvate carboxylase in *Yarrowia lipolytica*. *Appl. Microbiol. Biotechnol.* **98**, 2003–2013 (2014).

28. Campbell, B. *et al.* Pharmacokinetics, safety, and effects on exercise performance of L-arginine α-ketoglutarate in trained adult men. *Nutrition* **22**, 872–881 (2006).

29. Cynober, L.A. The use of alpha-ketoglutarate salts in clinical nutrition and metabolic care. *Curr. Opin. Clin. Nutr. Metab. Care* **2**, 33–37 (1999).

30. Yin, X., Madzak, C., Du, G., Zhou, J., & Chen, J. Enhanced alpha-ketoglutaric acid production in *Yarrowia lipolytica* WSH-Z06 by regulation of the pyruvate carboxylation pathway. *Appl. Microbiol. Biotechnol.* **96**, 1527–1537 (2012).

31. Weissbrodt, E. *et al.* in DD Patent No. 267999 (1989).
32. Yua, Z., Du, G., Zhou, J., & Chen, J. Enhanced α-ketoglutaric acid production in *Yarrowia lipolytica* WSH-Z06 by an improved integrated fed-batch strategy. *Bioresour. Technol.* **114**, 597–602 (2012).
33. Otto, C., Yovkova, V., Aurich, A., Mauersberger, S., & Barth, G. Variation of the by-product spectrum during α-ketoglutaric acid production from raw glycerol by overexpression of fumarase and pyruvate carboxylase genes in *Yarrowia lipolytica. Appl. Microbiol. Biotechnol.* **95**, 905–917 (2012).
34. Zhou, J., Yin, X., Madzak, C., Du, G., & Chen, J. Enhanced α-ketoglutarate production in *Yarrowia lipolytica* WSH-Z06 by alteration of the acetyl-CoA metabolism. *J. Biotechnol.* **161**, 257–264 (2012).
35. Guo, H., Madzak, C., Du, G., Zhou, J., & Chen, J. Effects of pyruvate dehydrogenase subunits overexpression on the α-ketoglutarate production in *Yarrowia lipolytica* WSH-Z06. *Appl. Microbiol. Biotechnol.* **98**, 7003–7012 (2013).
36. Jambunathan, P., & Zhang, K. Novel pathways and products from 2-keto acids. *Curr. Opin. Biotechnol.* **29**, 1–7 (2014).
37. Peters-Wendisch, P.G. *et al.* Pyruvate carboxylase is a major bottleneck for glutamate and lysine production by *Corynebacterium glutamicum. J. Mol. Microbiol. Biotechnol.* **3**, 295–300 (2001).
38. Sawada, K., Zen-in, S., Wada, M., & Yokota, A. Metabolic changes in a pyruvate kinase gene deletion mutant of *Corynebacterium glutamicum* ATCC 13032. *Metab. Eng.* **12**, 401–407 (2010).
39. Kim, J. *et al.* Effect of *odhA* overexpression and *odhA* antisense RNA expression on Tween-40-triggered glutamate production by *Corynebacterium glutamicum. Appl. Genet. Mol. Biotechnol.* **81**, 1097–1106 (2009).
40. Kim, J. *et al.* Requirement of *de novo* synthesis of the OdhI protein in penicillin-induced glutamate production by *Corynebacterium glutamicum. Appl. Genet. Mol. Biotechnol.* **86**, 911–920 (2010).
41. Liu, Q., Ouyang, S.-P., Kim, J., & Chen, G.-Q. The impact of PHB accumulation on L-glutamate production by recombinant *Corynebacterium glutamicum. J. Biotechnol.* **132**, 273–279 (2007).
42. Chinen, A., Kozlov, Y.I., Hara, Y., Izui, H., & Yasued, H. Innovative metabolic pathway design for efficient L-glutamate production by suppressing CO_2 emission. *J. Biosci. Bioeng.* **103**, 262–269 (2007).
43. Baynes, B.M., U.S. Patent No. 8133704 (2012).
44. Turk, S.C.H.J. *et al.* Metabolic engineering toward sustainable production of nylon-6. *ACS Synth. Biol.* **5**, 65–73 (2016).
45. Jansen, M.L., & Gulik, W.M.v. Towards large scale fermentative production of succinic acid. *Curr. Opin. Biotechnol.* **30**, 190–197 (2014).

46. Wang, W., Li, Z., Xie, J., & Ye, Q. Production of succinate by a *pflB ldhA* double mutant of *Escherichia coli* overexpressing malate dehydrogenase. *Bioproc. Biosyst. Eng.* **32**, 737–745 (2009).

47. Jantama, K. *et al.* Combining metabolic engineering and metabolic evolution to develop nonrecombinant strains of *Escherichia coli* C that produce succinate and malate. *Biotechnol. Bioeng.* **99**, 1140–1153 (2007).

48. Jantama, K. *et al.* Eliminating side products and increasing succinate yields in engineered strains of *Escherichia coli* C. *Biotechnol. Bioeng.* **101**, 881–893 (2008).

49. Balzera, G.J., Thakkerc, C., Bennettc, G.N., & San, K.-Y. Metabolic engineering of *Escherichia coli* to minimize byproduct formate and improving succinate productivity through increasing NADH availability by heterologous expression of NAD$^+$-dependent formate dehydrogenase. *Metab. Eng.* **20**, 1–8 (2013).

50. Lin, H., Bennett, G.N., & San, K.-Y. Fed-batch culture of a metabolically engineered *Escherichia coli* strain designed for high-level succinate production and yield under aerobic conditions. *Biotechnol. Bioeng.* **90**, 775–779 (2005).

51. Raab, A.M., Gebhardt, G., Bolotina, N., Weuster-Botz, D., & Lang, C. Metabolic engineering of *Saccharomyces cerevisiae* for the biotechnological production of succinic acid. *Metab. Eng.* **12**, 518–525 (2010).

52. Yan, D. *et al.* Construction of reductive pathway in *Saccharomyces cerevisiae* for effective succinic acid fermentation at low pH value. *Bioresour. Technol.* **156**, 232–239 (2014).

53. Yim, H. *et al.* Metabolic engineering of *Escherichia coli* for direct production of 1,4-butanediol. *Nat. Chem. Biol.* **7**, 445–452 (2011).

54. Barton, N.R. *et al.* An integrated biotechnology platform for developing sustainable chemical processes. *J. Ind. Microbiol. Biotechnol.* **42**, 349–360 (2015).

55. Engel, C.A.R., Straathof, A.J.J., Zijlmans, T.W., Gulik, W.M.v., & Wielen, L.A.M.v.d. Fumaric acid production by fermentation. *Appl. Microbiol. Biotechnol.* **78**, 379–389 (2008).

56. Xu, Q., Li, S., Huang, H., & Wen, J. Key technologies for the industrial production of fumaric acid by fermentation. *Biotechnol. Adv.* **30**, 1685–1696 (2012).

57. Goldberg, I., Rokem, J.S., & Pines, O. Organic acids: Old metabolites, new themes. *J. Chem. Technol. Biotechnol.* **81**, 1601–1611 (2006).

58. Zhou, Y. *et al.* Production of fumaric acid from biodiesel-derived crude glycerol by *Rhizopus arrhizus*. *Bioresour. Technol.* **163**, 48–53 (2014).

59. Zhang, B., Skory, C.D., & Yang, S.-T. Metabolic engineering of *Rhizopus oryzae*: Effects of overexpressing *pyc* and *pepc* genes on fumaric acid biosynthesis from glucose. *Metab. Eng.* **14**, 512–520 (2012).
60. Zhang, B., & Yang, S.-T. Metabolic engineering of *Rhizopus oryzae*: Effects of overexpressing *fumR* gene on cell growth and fumaric acid biosynthesis from glucose. *Process Biochem.* **47**, 2159–2165 (2012).
61. Li, N. *et al.* Engineering *Escherichia coli* for fumaric acid production from glycerol. *Bioresour. Technol.* **174**, 81–87 (2014).
62. Song, C.W., Kim, D.I., Choi, S., Jang, J.W., & Lee, S.Y. Metabolic engineering of *Escherichia coli* for the production of fumaric acid. *Biotechnol. Bioeng.* **110**, 2025–2034 (2013).
63. Xu, G., Chen, X., Liua, L., & Jiang, L. Fumaric acid production in *Saccharomyces cerevisiae* by simultaneous use of oxidative and reductive routes. *Bioresour. Technol.* **148**, 91–96 (2013).
64. Chen, X., Zhu, P., & Liu, L. Modular optimization of multi-gene pathways for fumarate production. *Metab. Eng.* **33**, 76–85 (2016).
65. Chen, X., Dong, X., Wang, Y., Zhao, Z., & Liu, L. Mitochondrial engineering of the TCA cycle for fumarate production. *Metab. Eng.* **31**, 62–73 (2015).
66. Könst, P.M., Franssen, M.C.R., Scott, E.L., & Sanders, J.P.M. A study on the applicability of L-aspartate alpha-decarboxylase in the biobased production of nitrogen containing chemicals. *Green Chem.* **11**, 1646–1652 (2009).
67. Steunenberg, P., Könst, P.M., Scott, E.L., Franssen, M.C.R., & Zuilhof, H. Polymerisation of beta-alanine through catalytic ester-amide exchange. *Eur. Polym. J.* **49**, 1773–1781 (2013).
68. Song, C.W., Lee, J., Ko, Y.-S., & Lee, S.Y. Metabolic engineering of *Escherichia coli* for the production of 3-aminopropionic acid. *Metab. Eng.* **30**, 121–129 (2015).
69. Song, C.W., Kim, J.W., Cho, I.J., & Lee, S.Y. Metabolic engineering of *Escherichia coli* for the production of 3-hydroxypropionic acid and malonic acid through β-alanine route. *ACS Synth. Biol.* **5**(11), 1256–1263 (2016), doi:10.1021/acssunbio.1026b00007.
70. Chi, Z., Wang, Z.-P., Wang, G.-Y., Khan, I., & Chi, Z.-M. Microbial biosynthesis and secretion of L-malic acid and its applications. *Crit. Rev. Biotechnol.* **36**, 99–107 (2016).
71. Alonso, S., Rendueles, M., & Díaz, M. Microbial production of specialty organic acids from renewable and waste materials. *Crit. Rev. Biotechnol.* **35**, 497–513 (2015).

72. Battat, E., Peleg, Y., Bercovitz, A., Rokem, J.S., & Goldberg, I. Optimization of L-malic acid production by *Aspergillus flavus* in a stirred fermentor. *Biotechnol. Adv.* **37**, 1108–1116 (1991).

73. Moon, S.Y., Hong, S.H., Kim, T.Y., & Lee, S.Y. Metabolic engineering of *Escherichia coli* for the production of malic acid. *Biochem. Eng. J.* **40**, 312–320 (2008).

74. Zhang, X., Wang, X., Shanmugam, K.T., & Ingram, L.O. L-Malate production by metabolically engineered *Escherichia coli. Appl. Environ. Microbiol.* **77**, 427–434 (2011).

75. Zelle, R.M. *et al.* Malic acid production by *Saccharomyces cerevisiae*: Engineering of pyruvate carboxylation, oxaloacetate reduction, and malate export. *Appl. Environ. Microbiol.* **82**, 2766–2777 (2016).

76. Zou, X., Zhou, Y., & Yang, S.-T. Production of polymalic acid and malic acid by *Aureobasidium pullulans* fermentation and acid hydrolysis. *Biotechnol. Bioeng.* **110**, 2105–2113 (2013).

77. Park, S., Chang, K.S., Jin, E., Pack, S.P., & Lee, J. Oxaloacetate and malate production in engineered Escherichia coli by expression of codon-optimized phosphoenolpyruvate carboxylase2 gene from *Dunaliella salina. Bioprocess. Biosyst. Eng.* **36**, 127–131 (2012).

78. Park, S., Hong, S., Pack, S.P., & Lee, J. High activity and stability of codon-optimized phosphoenolpyruvate carboxylase from *Photobacterium profundum* SS9 at low temperatures and its application for *in vitro* production of oxaloacetate. *Bioprocess. Biosyst. Eng.* **37**, 331–335 (2014).

79. Borodina, I. *et al.* Establishing a synthetic pathway for high-level production of 3-hydroxypropionic acid in *Saccharomyces cerevisiae* via β-alanine. *Metab. Eng.* **27**, 57–64 (2014).

80. Ikram-ula, H., Alia, S., Qadeerb, M.A., & Iqbal, J. Citric acid production by selected mutants of *Aspergillus niger* from cane molasses. *Bioresour. Technol.* **93**, 125–130 (2004).

81. Bassalo, M.C. *et al.* Rapid and efficient one-step metabolic pathway integration in *E. coli. ACS Synth. Biol.*, **5**(7), 561–568 (2016), doi:10.1021/acssynbio.1025b00187.

82. Shi, S., Liang, Y., Zhang, M.M., Ang, E.L., & Zhao, H. A highly efficient single-step, markerless strategy for multi-copy chromosomal integration of large biochemical pathways in *Saccharomyces cerevisiae. Metab. Eng.* **33**, 19–27 (2016).

83. Tyo, K.E.J., Ajikumar, P.K., & Stephanopoulos, G. Stabilized gene duplication enables long-term selection-free heterologous pathway expression. *Nat. Biotechnol.* **27**, 760–767 (2009).

84. Liu, H., & Lu, T. Autonomous production of 1,4-butanediol via a *de novo* biosynthesis pathway in engineered *Escherichia coli*. *Metab. Eng.* **29**, 135–141 (2015).

85. Li, H. *et al.* Integrated electromicrobial conversion of CO_2 to higher alcohols. *Science* **335**, 1596 (2012).

86. Choi, K.-Y., Wernick, D.G., Tat, C.A., & Liao, J.C. Consolidated conversion of protein waste into biofuels and ammonia using *Bacillus subtilis*. *Metab. Eng.* **23**, 53–61 (2014).

87. Müller, J.E.N. *et al.* Engineering *Escherichia coli* for methanol conversion. *Metab. Eng.* **28**, 190–201 (2015).

88. Conrado, R.J., & Gonzalez, R. Chemistry envisioning the bioconversion of methane to liquid fuels. *Science* **343**, 621–623 (2014).

89. Haynes, C.A., & Gonzalez, R. Rethinking biological activation of methane and conversion to liquid fuels. *Nat. Chem. Biol.* **10**, 331–339 (2014).

90. Royce, L.A. *et al.* Evolution for exogenous octanoic acid tolerance improves carboxylic acid production and membrane integrity. *Metab. Eng.* **29**, 180–188 (2015).

91. Kildegaard, K.R. *et al.* Evolution reveals a glutathione-dependent mechanism of 3-hydroxypropionic acid tolerance. *Metab. Eng.* **26**, 57–66 (2014).

92. Rai, A., & Saito, K. Omics data input for metabolic modeling. *Curr. Opin. Biotechnol.* **37**, 127–134 (2016).

93. Cardona, C., Weisenhorn, P., Henry, C., & Gilbert, J.A. Network-based metabolic analysis and microbial community modeling. *Curr. Opin. Microbiol.* **31**, 124–131 (2016).

94. Karr, J.R. *et al.* A whole-cell computational model predicts phenotype from genotype. *Cell* **150**, 389–401 (2012).

Chapter 3

Amino Acid Biosynthesis and Its Metabolic Engineering Applications

Yi-Xin Huo

College of Life Science,
Beijing Institute of Technology, Beijing, China
huoyixin@bit.edu.cn

3.1. Introduction of Amino Acids

Amino acids are defined as organic substances containing both amino and acid groups. All amino acids, except for glycine, have an asymmetric carbon and exhibit optical activities. The absolute configuration of an amino acid (L- or D-isomers) is defined with reference to glyceraldehydes. Amino acids could be grouped based on their polarity and side-chain group type (aliphatic or aromatic, containing hydroxyl or sulfur, etc.).

Except for proline, all universal proteinogenic amino acids are α-amino acids, which have a primary amino group and a carboxyl group linked to the α-carbon atom. On the other hand, if an amino group links to the β-carbon atom, the amino acid is defined as β-amino acids (e.g., taurine and β-alanine). Additionally, in some unusual proteins, special amino acids are post-translationally modified.[1] The chemical name, code, relative molecular mass, chemical structure, molecular formula, and optical activities of the 20 universal amino acids are summarized in Table 3.1.

Table 3.1. Chart of amino acids.

His Histidine H 155.16 137.14 -38.5 +11.8 $C_6H_9N_3O_2$	Arg Arginine R 174.20 156.19 +12.5 +27.6 $C_6H_{14}N_4O_2$	Lys Lysine K 146.19 128.17 +13.5 +26.0 $C_6H_{14}N_2O_2$	Asp Aspartic Acid D 133.10 115.09 +5.0 +25.4 $C_4H_7NO_4$	Glu Glutamic Acid E 147.13 129.11 +12.0 +31.8 $C_5H_9NO_4$
Ile Isoleucine I 131.18 113.16 +12.4 +39.5 $C_6H_{13}NO_2$	**Phe** Phenylalanine F 165.19 147.18 -34.5 -4.5 $C_9H_{11}NO_2$	**Leu** Leucine L 131.17 113.16 -11.0 +16.0 $C_6H_{13}NO_2$	**Trp** Tryptophan W 204.23 186.21 -33.7 +2.8 $C_{11}H_{12}N_2O_2$	**Ala** Alanine A 89.09 71.08 +1.8 +14.6 $C_3H_7NO_2$
Met Methionine M 149.21 131.20 -10.0 +23.2 $C_5H_{11}NO_2S$	**Pro** Proline P 115.13 97.12 -86.2 -60.4 $C_5H_9NO_2$	**Val** Valine V 117.15 99.13 +5.6 +28.3 $C_5H_{11}NO_2$	**Cys** Cysteine C 121.16 103.14 -16.5 +6.5 $C_3H_7NO_2S$	**Asn** Asparagine N 132.12 114.10 -5.3 +33.2② $C_4H_8N_2O_3$
Gly Glycine G 75.07 57.05 — — $C_2H_5NO_2$	**Ser** Serine S 105.09 87.08 -7.5 +15.1 $C_3H_7NO_3$	**Gln** Glutamine Q 146.15 128.13 +6.3 +31.8① $C_5H_{10}N_2O_3$	**Tyr** Tyrosine Y 181.19 163.17 — -10.0 $C_9H_{11}NO_3$	**Thr** Threonine T 119.12 101.10 -28.5 -15.0 $C_4H_9NO_3$

1-Letter Amino Acid Code
Relative Molecular Mass
M_r-H_2O
$[\alpha]_D(H_2O)$
$[\alpha]_D(5 mol/L HCl$
or ①1 mol/L HCl
or ②3 mol/L HCl)

Asp — 3-Letter Amino Acid Code
Aspartic Acid — Chemical Name
D
133.10
115.09 — Chemical Structure
+5.0
+25.4
$C_4H_7NO_4$ — Molecular Formula

■ Basic
■ Non-polar (hydrophobic)
■ Polar, uncharged
■ Acidic

The basic, non-polar, polar, and acidic groups of amino acids are marked with different colors. Each amino acid has its own unique biochemical properties and functions, determined by its side chain.[2–5]

The discovery, quantitative measurement, and qualitative analysis of amino acids were facilitated by analytical methods such as crystallization from protein hydrolysates, separation of corresponding esters by vacuum distillation, and precipitation by sulfonic acid derivatives.

DL-Alanine is the first chemically synthesized amino acid, followed by glycine and DL-methionine. However, amino acids produced through chemical approaches are racemic, whereas the pure L- or D-form amino acid is what customers demanded. All L-amino acids were manufactured by separation from the chemically synthesized racemic mixtures or by isolation from protein hydrolyzates until L-glutamic acid fermentation process was developed in mid-1950s using glutamic acid bacteria (e.g., *Corynebacterium glutamicum, Brevibacterium flavum*) as a microbial factory. Since then, the strategies and methods of producing L-amino acids have changed extensively.[6]

Nowadays, most amino acids (except for glycine, L-methionine, L-cysteine, and L-serine) are commercially manufactured by the fermentation process. Alternatively, *in vitro* enzymatic processes are utilized to produce L-alanine, L-aspartic acid, L-cysteine, L-serine, L-tryptophan, L-lysine, L-phenylalanine, and some D-amino acids. When racemic mixture is acceptable, chemical synthesis is still the dominant process for producing glycine, DL-alanine, DL-methionine, and DL-cysteine.

At physiological conditions, amino acids are generally stable in aqueous solution except for glutamine and cysteine. The solubility of amino acids is summarized in Table 3.2. An amino acid might precipitate in the reactors after its titer exceeds the corresponding solubility, facilitating the separation, and purification process.

Table 3.2.   Solubility of amino acids.

Amino acid	Solubility in water, g/L	
	25°C	50°C
Gly L-	250	391
Ala L-	166.5	217.9
Val L-	88.5	96.2
Leu L-	24.26	28.87
Ile L-	41.2	48.2
Pro L-	1623	2067
Phe L-	29.6	44.3
Trp L-	11.4	17.1
Met DL-	33.81	60.70
Asp L-	5.0	12.0
Glu L-	8.64	21.86

(Continued)

Table 3.2. *(Continued)*

Amino acid	Solubility in water, g/L	
	25°C	50°C
Tyr L-	0.453	1.052
His L-	41.9	–
Arg L-	the satd aq soln contains 15% (w/w) at 21°C	
Ser L-	soluble	
Cys L-	freely soluble	
Thr L-	freely soluble	
Lys L-	freely soluble	

3.2. Application and Market Volume of Amino Acids

The worldwide total annual consumption of amino acids has increased significantly during the past decade, from two million tons in 2005 to over four million tons in 2015. The biggest share of the amino acid market is held by monosodium glutamate, L-lysine, DL-methionine, and L-threonine. Monosodium glutamate, used as flavor enhancer for food, has an annual production volume over 1.5 million tons worldwide.[7] The most important applications of other amino acids are feed additives. The percentage of crude protein and the limiting amino acids in common feedstuffs are summarized in Tables 3.3.

L-Lysine is the most important feed additive and has an annual fermentative production about 0.75 million tons. It is an essential amino acid required for humans and animals, which has to be supplied in food and feed. Methionine, the essential sulfur-containing amino acid, is the most limiting amino acid in conventional poultry feed[8–10] since its concentration in many plant-based diets are low.[11] Therefore, methionine or its hydroxyl analog 2-hydroxy-4-(methylthio) butanoic acid[12] needs to be routinely added to animal feed. The market volume of methionine and its analog are at least 0.5 million tons per year. L-Threonine is an essential amino acid almost exclusively used as a feed additive and has an annual world market volume of about 70,000 tons. L-Threonine is primarily added to pig and poultry diets.

Table 3.3. Limiting amino acids of common feedstuffs for pigs and chicken.

Pig			Chicken	
Second	First	Ingredient/Crude protein, %	First	Second
Trp	Lys	Maize/8.9	Lys	Trp
Thr	Lys	Sorghum/9.5	Lys	Arg
—	Lys	Barley/12.6	Lys	Met
Trp	Lys	Wheat/12.6	Lys	Thr
Thr	Met	Soybean meal/46.2	Met	Thr
—	—	Fish meal/64.3	Arg	—
—	Met	Rapeseed meal/35.3	Lys	Arg
—	Lys	Peanut meal/47.4	Met	Lys
—	Lys	Sunflower seed meal/31.7	Met	Thr
—	Lys	Meat and bone meal/48.6	Trp	Met
Thr	Lys	Cottonseed meal/64.3	Lys	Met
Trp	Lys	Com gluten meal/63.6	Lys	Trp

Pig fed on maize, then the feedstuffs should add lysine first and tryptophan second.
"—" means no addition.

The market values of L-lysine and L-threonine have triggered intensive research on their biosynthetic pathways as well as their regulation, which resulted in the successfully engineered L-lysine and L-threonine production strain. Nevertheless, most of the methionine is supplied as a chemically produced racemate due to the lack of successful fermentation process. The development of economically feasible biosynthesis process for L-methionine has attracted widespread attention.

In the plant-based animal feed market, L-tryptophan is the fourth limiting and in-demand amino acid after L-lysine, D/L-methionine, and L-threonine.[13] Industrial production of L-tryptophan, an aromatic amino acid, is mostly for feed and pharmaceutical purposes, and the world requirement was only 3,000 tons in 2005 due to its high price (as high as $22–24/kg). This relatively small volume of L-tryptophan could not be satisfied by hydrolyzing protein biomass since tryptophan is not stable during the acidic hydrolytic processes. Same as L-tryptophan, L-phenylalanine is termed an essential aromatic amino acid for man and most livestock (except ruminants). The major application of L-phenylalanine is to produce aspartame (*N*-l-α-aspartyl-l-phenylalanine 1-methyl ester), which is about 200× as sweet as sucrose[13] and serves as a low-caloric sweetener in diet drinks or food.[14] L-Aspartame has a total annual market volume of about 18,000 tons. L-Tyrosine, the third aromatic amino acid, could be converted from L-phenylalanine via an enzymatic hydroxylation process. L-Tyrosine is not an essential amino acid in the presence of L-phenylalanine, and its demand is rather low (<200 tons per year). The above three aromatic amino acids also serve as important building blocks for chemically synthesizing pharmaceutically active compounds.

Leucine, isoleucine, and valine are branched-chain amino acids (BCAAs), which are the most hydrophobic of the amino acids (Table 3.1) due to their unsubstituted aliphatic chains and branched alkyl groups. BCAAs are essential for human and other vertebrates. L-aline is one of the top five most limiting amino acids in meal diets for livestock.[15] About 500 and 200 tons of L-valine are produced annually by fermentation[16] and by enzyme membrane reactor methods.[13] L-isoleucine is produced about 400 tons per year,[17] utilizing *Corynebacterium glutamicum* and *Escherichia coli* as the best microbial factories.[18] It is especially needed for pigs during both early and late feeding stages.[19] The market volume of total amino acid increases by about 10% annually[18] as the growing demand of BCAAs is one of the driving forces. The price of the fermentation produced BCAAs is expected to keep decreasing, which would expand the applications of BCAAs in both food and feed market.[20]

About 300 tons of L-serine are produced annually,[21] which is used as the compound of infusion solutions in the pharmaceutical industry, serves as a moisturizing agent in skin lotions in the cosmetics industry, and presents as an additive to drinks in the food industry. Besides the above applications, a substantial amount of L-serine is enzymatically converted to L-tryptophan in the presence of indole,[22] although most L-tryptophan in the market were produced by fermentation using *E. coli*, *C. glutamicum*, or *B. subtilis*.[23]

About 22,000 tons of the sweet glycine are produced annually,[21] which is used as a food and feed additive to ameliorate the flavor and taste of vinegar, fruit juices, and salted vegetables. In the fertilizer industry, glycine serves as a solvent for removing CO_2.

De novo L-arginine productions occur in many plants, fungi, and prokaryotes. L-Arginine is a conditionally essential amino acid for humans, which become deficient under conditions of increased demand such as growth or tissue repair.[24,25] Furthermore, it is reported that L-arginine has an immune-supportive effect[26] and is the source of nitrogen oxide, a vasodilatory messenger[27] involved in cellular communications.[28] However, the market of L-arginine is fairly small.

Taken together, amino acids are widely used as food additives, feed supplements, cosmetics, pharmaceuticals, and polymer precursors. In fact, most of the amino acids have the potential to be utilized in the food industry as a taste adjuster. The taste profile of the amino acids is summarized in Figure 3.1. Novel economically feasible fermentation processes are required to meet the demand of the global amino acid market, which increases by 5–7% annually.[13]

3.3. Economic Aspects of the Biosynthesis of Amino Acids

Usually the price of a specific amino acid decreases when its market size and production increases. The relationship between the production capacity and the selling prices is summarized in Figure 3.2. This indicates the need for process optimization and sustainable strain development. While the market grew, the size of the bioreactors and facilities increased stepwise. Nowadays, 50–500 m³ bioreactors are standard size for amino acid production (depending on the product). Hence, all production processes have to be optimized to work perfectly under conditions in reactor volumes up to 500 m³.

It's hard to receive monopoly profit for common amino acids such as L-lysine and L-threonine although the costs of raw materials and fermentation already have been minimized and well controlled in every manufacture. Therefore, the cost of downstream processing will play a significant role in determining the selling price of an amino acid. To lower the downstream processing cost, different products with different purities, concentrations, and by-products are prepared to satisfy the different requirements of different customers. For example, lysine products could be categorized into liquid lysine (50% purity), granulated lysine sulfate (40–50% purity),

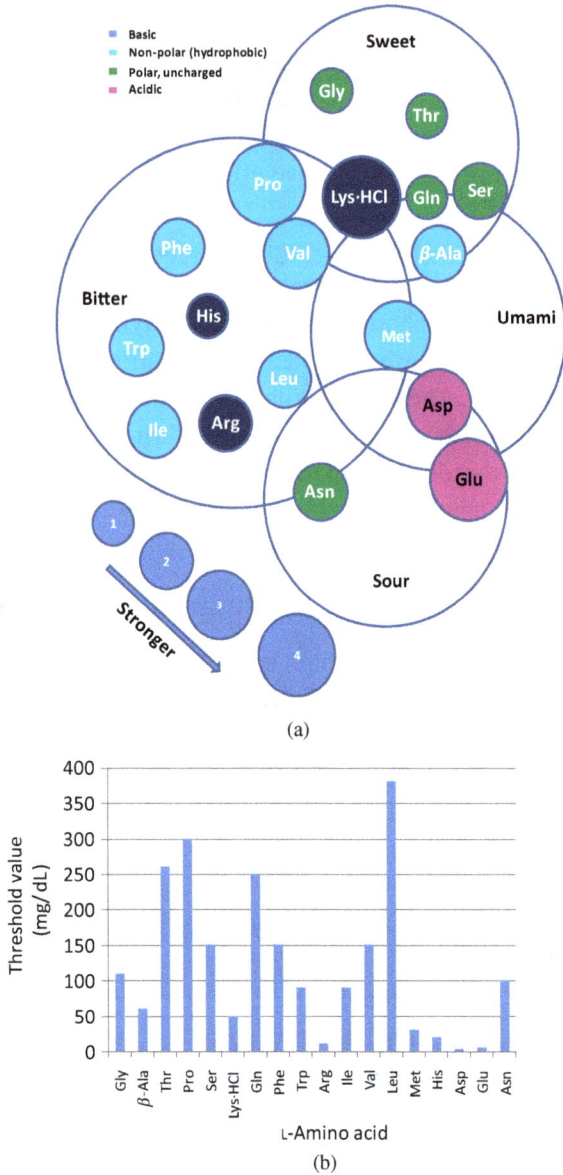

Figure 3.1. Taste profiles of L-amino acids. (a) The taste of each amino acid in the group of sweet, bitter, umami, and sour. For example, glycine tasted sweet, methionine tasted more bitter than umami, lysine (with HCl) tasted more sweet than bitter and umami. The taste intensity of each amino acid is expressed as follows: (1) detectable, (2) strong, (3) stronger, (4) strongest. (b) Threshold concentration of each L-amino acids to be tasted by human being.

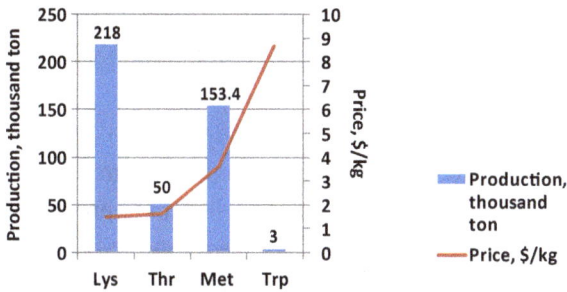

Figure 3.2. Global production of four amino acids in 2015. Among them, lysine has the highest annual production and the lowest price per kg, while tryptophan has the lowest annual production and the highest price per kg.

and liquid lysine sulfate (20–30% purity). The granulated product contains the entire fermentation broth without the separation of microbial biomass, which minimized the separation cost and provides additional nutritional value.[29] However, it could only be sold as feed additive and is not allowed to enter the food, pharmaceutical, or cosmetic market.

In contrast to the price competition in the feed market, more profit margins could be obtained by producing pharmaceutical intermediate or amino acid derivatives with less competition. Not surprisingly, prices for pharma-grade amino acids such as L-tryptophan are still very high (as high as $100/kg).[13] Unlike all other amino acids produced at a large scale, no fermentation process has been developed for methionine. The environmental pressure and the increasing prices of chemical precursors (acrolein, methylmercaptan, and hydrogen cyanide) have sparked the interest in the fermentative production of L-methionine. If successfully developed, the fermentation process will have the potential to dominate its production and market.

3.4. Amino Acid Biosynthesis Pathways and Their Regulations

The amino acids are essential to all living organisms, and they serve as the building blocks of our proteins and could be converted into a variety of physiologically important metabolites such as lipids, hormones, and amino sugars. The biosynthesis of amino acids from glucose has been well studied and is simplified and summarized in Figure 3.3. In this chapter, we only discuss the pathways upon which modern fermentation industry was established and would be further developed. The internal conversion among different amino acids is summarized as Figure 3.4.

The amino acids L-glutamate and L-glutamine are the storage pools of ammonia and serve as key nitrogen or amino residue donors for the synthesis of other amino acids, proteins, nucleic acids, and cell wall

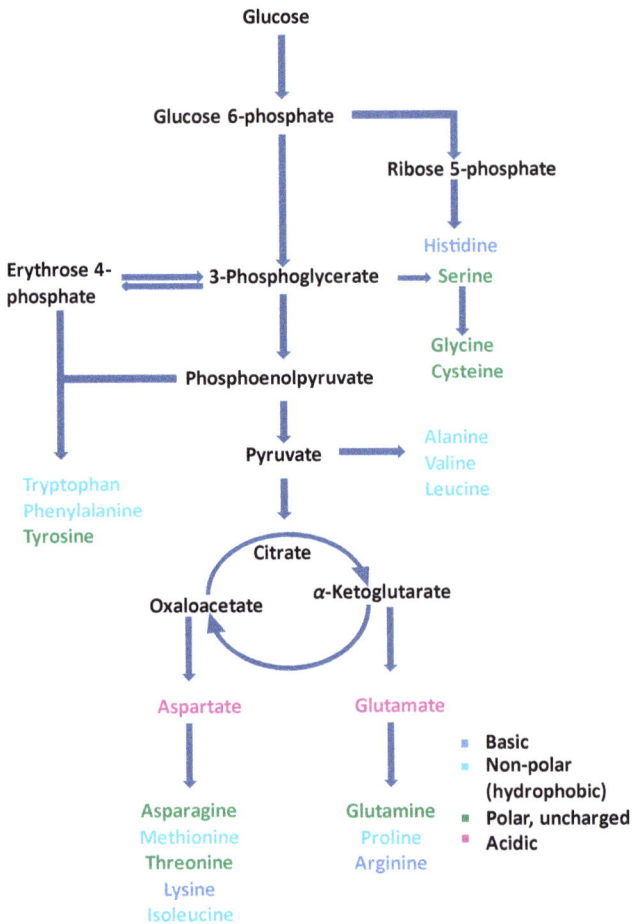

Figure 3.3. Overview of amino acid biosynthesis using glucose as the carbon source.

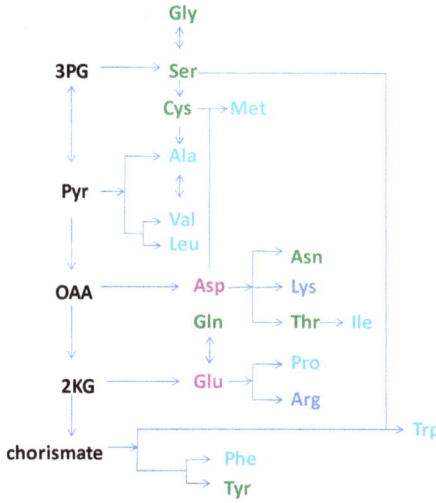

Figure 3.4. Conversion of L-Amino Acids. 3PG: 3-phospho-D-glycerate, Pyr: pyruvate, OAA: oxaloacetate, 2KG: 2-oxoglutarate, Arrows: multiple steps. The color of amino acids is marked as in Figure 3.3.

components. L-Glutamate could be converted into glutamine by a one-step reaction catalyzed by glutamine synthetase (GS) while L-glutamate could be converted from its precursor α-ketoglutarate in reactions catalyzed by glutamate dehydrogenase (GDH) or glutamine α-oxoglutarate amidotransferase (GOGAT), depending on the environmental ammonium concentration and other regulatory factors.[30]

L-Lysine could be synthesized through two completely different routes, either from aspartate via the diaminopimelate route or from α-oxoglutarate and acetyl-CoA via the α-aminoadipate route. Furthermore, the details of α-aminoadipate route are different in bacterium such as *Thermus thermophilus* and in archaea such as *Thermoproteus neutrophilus*.[31]

In most organisms, biosynthesis pathways of BCAAs (all in their L-forms) form a super pathway. L-Valine and L-isoleucine are synthesized in parallel pathways sharing the same enzymes, while L-leucine is synthesized through a specific series of reactions but shares common precursors as L-valine. The metabolite flux through these pathways was strictly controlled by a complicated regulatory network, in which both gene expressions and enzyme activities were regulated. Furthermore, BCAA has similar structures and could mimic each other in binding to the catalytic or regulatory sites of the related enzymes to create the feedback inhibition effect. For example, the presence of high concentration of L-isoleucine or its precursor 2-ketobutyrate could inhibit the biosynthesis of L-valine and L-leucine.[32]

L-Tyrosine is produced from ammonia, pyruvate, and phenol via the enzyme tyrosine phenol-lyase (EC 4.1.99.2) in microbial production strains such as *Erwinia herbicola*.[33] Although this pathway has the potential to achieve high production titer and yield, the majority of tryptophan is industrially produced by microbial fermentations with *Corynebacteria* and *E. coli* via shikimate pathway.[14,16,23,34–36]

L-Phenylalanine can be obtained by chemical, enzymatic, or microbial processes. Enzymatic reaction relies on either the amination of *trans*-cinnamic acid or reductive amination of phenylpyruvate.[37–39] Microbial production of L-phenylalanine from phenylpyruvate and aspartate could be achieved in recombinant *E. coli* cells with overexpressed aminotransferases and PEP carboxykinase.[37] Generally, recombinant strains of *C. glutamicum* and *E. coli* are utilized for the fermentative production of phenylalanine.[14,16,23,36,40,41]

In recombinant *E. coli*, production strains that overproduce tryptophanase (gene *tnaA*), L-tryptophan was produced from indole, pyruvate, and ammonia.[42] Alternatively, in recombinant *E. coli* production strains that overproduce tryptophan synthase, L-tryptophan was produced from indole and L-serine.[43,44]

Although most microorganisms could synthesize L-methionine, no fermentative production process has been established for it yet. In both *E. coli* and *C. glutamicum*, the biosynthesis of L-methionine was regulated on the transcriptional level. An activator named MetR and a repressor named MetJ were identified in *E. coli* while a transcriptional factor named McbR was demonstrated as the master regulator in *C. glutamicum*. In both organisms, homoserine acyltransferase is the first enzyme specific to L-methionine biosynthesis and is the regulatory target of the feedback inhibition. In *C. glutamicum*, the activities of other enzymes such as cystathionine-γ-synthase, cystathionine-β-lyase, and O-acetyl-homoserinesulfhydrylase are also feedback inhibited by the accumulation of L-methionine.

3.5. DE0057 Production Strain

C. glutamicum is the most important organism for industrial amino acid production.[45,46] It is a gram-positive, rod-like, aerobic, and glutamate producing soil bacterium, and was isolated in 1956.[12,47] This strain drew much attention and was investigated to produce another amino acid stemming from tricarboxylic acid (TCA) cycle: L-lysine. Nowadays, *C. glutamicum* is used to produce several million tons of amino acids annually, in particular the flavor enhancer L-glutamate (2,200,000 tons/year) and the feed additive L-lysine (1,500,000 tons/year). For instance, the global production of lysine from 2004 to 2015 is summarized in Figure 3.5.

Glutamate production by *C. glutamicum* could be increased by many triggers such as depletion of biotin, addition of detergent or β-lactam antibiotics, and addition of ethambutol or cerulenin. However, the industrial production of L-glutamate and L-lysine were achieved through mutagenesis and screening. For example, Nakazawa *et al.*[48] obtained mutant strains of *C. glutamicum* that showed 10 to 1,000 times lower α-ketoglutarate dehydrogenase complex (ODHC) activity than the wild-type strain after treating the strains with the mutagen *N*-methyl-*N*-nitro-*N*-nitrosoguanidine. After the glutamate production by these mutants were investigated, results showed that the mutants could produce high amounts of glutamate in the

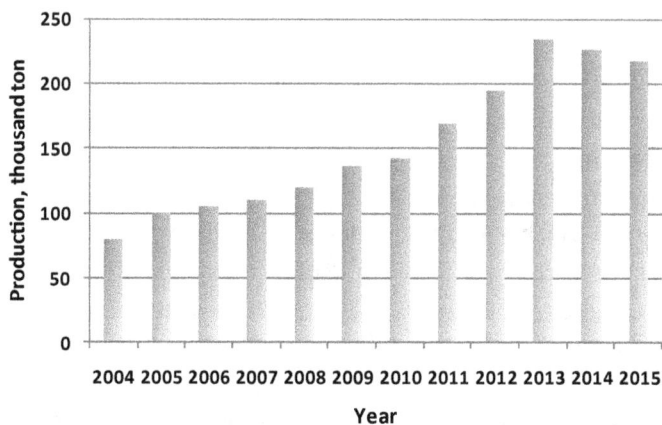

Figure 3.5.   Global production of lysine from 2004 to 2015.

presence of excess biotin. The carbon and nitrogen flux in metabolic pathways change dramatically after glutamate overproduction. Recombinant *E. coli* is the only other species used for lysine production.[49,50]

The traditional strategy for screening amino acid overproducer was to take the advantage of toxic analogs. For example, *S*-(2-aminoethyl) cysteine was utilized to screen for L-lysine feedback resistant strains.[51] Most lysine overproduction strains, therefore contain point mutations in the aspartokinase gene, encoding an enzyme as the target of the feedback inhibition by lysine and threonine,[52,53] *C. glutamicum*,[54] and *Serratia marcescens*[55] mutants that overproduce L-valine were isolated as resistant to BCAA analogs such as norvaline or 2-aminobutyricacid. In these strains, deregulation of α-acetohydroxy acid synthetase (AHAS) was probably the reason for valine overproduction.[54] Along this pathway, several steps are rate-limiting and must be engineered to overcome the metabolic barriers. The strains were first derived particularly as mutants resistant to analogs of threonine and isoleucine-like hydroxynorvaline and methylthreonine, respectively. Breeding of *C. glutamicum* leucine producers began as it did for other BCAAs by chemical mutagenesis and isolation of analog-resistant mutants. Mutants resistant to 2-azaleucine and 3-hydroxyleucine possessed isopropylmalate synthase and AHAS resistant to inhibition by leucine and by all three BCAAs, respectively.[54,56]

In an L-tyrosine producer named *C. glutamicum* (*B. lactofermentum*), the overexpression of *aroL* gene increased shikimate kinase activity by 10-fold in cell-free extracts and therefore increased L-tyrosine productivity by 25%.[36] The gene-dosage effect alone is sufficient to (partially) overcome regulations and limitations and to improve yields.

Experiments were performed to achieve L-methionine producer through similar mutation and screen strategy. Mutations in *metJ* or *metK* were screened out by using norleucine or ethionine as methionine analogs.[57] Feedback-resistant MetA alleles were obtained by selecting for α-methylmethionine resistance. In another early attempt, a *C. glutamicum* mutant resistant to the analogs ethionine, selenomethionine, and methionine hydroxamate was isolated and produced 2 g/L methionine.[58] The expression of methionine biosynthesis encoding genes was de-repressed in this strain. Mutant *E. coli* strains resistant to the threonine analog

α-amino-β-hydroxy valeric acid and ethionine or norleucine also produced up to 2 g/L methionine.[59] Mondal and colleagues have reported on
Brevibacterium heali strains that produce up to 25.5 g/L methionine.[60]
However, none of the analog-resistant strains has permitted the implementation of a commercially viable L-methionine production process.[61]

3.6. Novel Strategies to Overexpress Amino Acids

Today, systems-oriented approaches such as transcriptome, fluxome, and
metabolome were used to investigate cellular physiology and metabolism,
providing a novel powerful platform for understanding the amino acid
overproducing microorganisms and engineering superior production
strains. Improved precursor supply could be provided when checkpoints
for carbon flux and competing pathways were removed.

To remove the inhibition of AHAS in *C. glutamicum*,[62] site-directed
mutagenesis of the regulatory subunit is needed. Amino acid alterations
within the conserved *N*-terminus of the regulatory subunits of *E. coli*
ilvH*[63] and *S. cinnamonensis ilvN*[64] resulted in the resistance to L-valine
inhibition. A mutant AHAS entirely resistant to inhibition by all three
BCAAs was obtained by altering three consecutive amino acids in the
N-terminus of *C. glutamicum ilvN*.[62] These results showed that only a
single binding (allosteric) site was present at the enzyme molecule, which
exhibits different affinities to the three BCAAs. An impressively high
yield of 0.38 g of L-valine per g glucose was obtained.[65] Notably, the
newly identified L-valine exporter (encoded byygaZH) and its positive
regulatory circuit with Lrp dramatically enhanced the production of
L-valine. The overall strategies for metabolic engineering of *E. coli* results
in a theoretical maximum yield of 0.65 g L-valine per g glucose.[66]

Aiba and coworkers used *E. coli* mutants with lesions in *trpR* and
tnaA (encoding tryptophanase) to produce 6.2 g/L of L-tryptophan in
27 h.[67] With additional resistances against 6-fluoro-Trp and 8-azaguanine,
an engineered strain accumulated more than 50 g/L of L-tryptophan in
91h. L-Tryptophan crystallized during the process after addition of non-
ionic detergents.[68] However, the L-tryptophan production yields toward
sugar (weight %) was only between 20% and 25%.[16,23]

Several microorganisms such as *B. subtilis, Klebsiella aerogenes*, and *S. cerevisiae* can use methionine as sole sulfur source, indicating that this compound can be efficiently converted into cysteine.[69–71] In the presence of thiosulfate, cysteine production in the medium by overproducing YdeD reaches 75 mg/L. The overproduction of YfiK also leads to cysteine accumulation out of the cell (120 mg/L), but only in a *cysE* mutant strain, which has no feedback inhibition by cysteine. Strains with serine acetyl-transferase resistant to feedback inhibition by cysteine produce high quantities of cysteine and cystine (200 mg/L).[72]

Utagawa[73] showed that mutations of *C. glutamicum* subsp. *flavum*, a natural producer of glutamic acid, could produce as much as 25.3 g/L arginine. The arginine-producing strain (up to 4.5 g/L) was also resistant to 6-azauracil, a condition resulting from a deficiency in the *upp* gene (coding for uracil pyrophosphorylase).[74] Up to 19.3 g/L arginine was produced in the presence of glucose as a C source.[75] Deletion of a serine dehydratase gene along with the prevention of folate synthesis enables a *C. glutamicum* strain to reach maximal specific productivities of 1.45 mmol L^{-1} h^{-1} L-serine with a titer more than 50 g/L. On the other hand, a new screening of methanol-utilizing bacteria[76] resulted in a *Methylobacterium* strain exhibiting an exceptional high glycine conversion of 93% in the presence of methanol, with final L-serine concentrations of 65 g/L. In this case, the maximal specific productivity was 1.45 mmol L^{-1} h^{-1}, while the volumetric productivity was about 1.4 g L^{-1} h^{-1}.

Several recent microbial studies have also sought to generate D/L-alanine through fermentation. For example, an *Arthrobacter* strain accumulated L-alanine up to 75.6 g/L (with 1.2 g/L D-alanine) in 120 h with a mass yield of over 50% from glucose.[77] It is reported that *E. coli* expressing the *Arthrobacter* sp. HAP1 L-alanine dehydrogenase enzyme accumulated 2.9 g/L D/L-alanine or 8.1 g/L D/L-alanine under aerobic or oxygen-limited conditions in shake flasks.[78]

Systems and synthetic biology,[79,80] which targets the whole microorganisms and inspired the developments achieved in the field, is also being increasingly applied to industrial microbiology. Systems biology approaches that rely on models could improve either energy efficiency or carbon conversion efficiency of the production route. It has been proved to be a new useful optimization framework to formulate strain engineering. A genetic optimization program designed *in silico* could increase L-arginine production yield by up to 9%[81] in *E. coli*. The value of new technologies has already been demonstrated in industrial organisms, for example by precisely controlling of *r*RNA synthesis[82] or by predicting metabolic capabilities of *E. coli*.[83,84]

In light of the new powerful tools and principles of metabolic engineering, the quest for targeted development of methionine-producing strains is strongly revived.[85–87] The most promising candidates for methionine production are *C. glutamicum* and *E. coli*.[88] Elementary flux mode analysis and extreme pathway analysis might shed new light on the process development.[89–91] Several genome-scale studies have been reported for the production of succinate[92] and polyhydroxybutyrate in yeast,[93] and the growth-related aspects in *Saccharomyces cerevisiae*[94,95] and *E. coli*.[95–98]

Taken together, the combination of new systems/synthetic biology strategies and metabolic pathway analysis methods could be very powerful and useful in rational strain development. It allows metabolic engineers not only to determine the overall capacity of a cellular system, but also to study the general effects of any genetic modification. Moreover, the potential economic efficiency of a process could be estimated or simulated based on the theoretical maximum yield.

3.7. Fermentation of the Amino Acids

Nowadays, the amino acid fermentation is arguably the most mature biological process. Strains utilized in the L-lysine, L-glutamic acid, and L-threonine industry could reach high yields (about 50%), productivities (over 2.5 g L^{-1} h^{-1}), and titer (over 100 g/L). The results of fed-batch fermentation of other amino acids are relatively low, but are still promising.

Under fed-batch conditions, the best L-leucine producer strains produce more than 24 g/L, which is the solubility limit of L-leucine in the fermentation broth. A high molar product yield (0.30 mol L-leucine per mol glucose), and a promising volumetric productivity (4.3 mmol L^{-1} h^{-1}) were achieved in a defined minimal medium using a plasmid-free strain. The production titer, yield, and the precipitation of L-leucine in the bioreactor make this process useful for industrial application. Notably, the production of L-valine was greatly increased by a newly identified L-valine exporter (YgaZH) and its positive regulatory circuit with Lrp protein.[99] The L-valine yield could reach 0.32 gram of L-valine per gram of carbon sources (0.42 mol L-valine per mol carbon sources) when total amount of raw materials in the substrate (including glucose, acetic acid, and D-pantothenic acid, etc.) was considered in the calculation.[100]

The preparation time limits the overall efficiency of L-threonine production process. For example, *E. coli* B-3996 had an average volumetric L-threonine productivity of 1.77 g L^{-1} h^{-1} in a 36 h cultivation process.[101] However, if a 10-h preparation time was counted, the process productivity would only be 1.39 g L^{-1} h^{-1}. To solve this problem, a repeated fed-batch process[102] was developed to increase the process productivity by more than 20%, reaching 1.69 g L^{-1} h^{-1}.

The production titer of L-lysine could be increased to 8 g or 55 g/L from less than 1 g/L, respectively, through a *hom* or *lysC* mutations to the wild-type strain, indicating that both mutations are beneficial to the production. When combined, the two mutations had a synergistic effect on production (75 g/l). Further introduction of the beneficial *pyc* mutation could increase the titer to 80 g/L in a 27 h fermentation process.[103] This productivity (3.0 g L^{-1} h^{-1}) is so far the highest one reported among microbial L-lysine fed-batch fermentation processes.

L-Arginine production with AR6 strain could reach high yield (81.2 g L^{-1}) and productivity (0.91 g L^{-1} h^{-1}) using a mixture of glucose and sucrose as raw material.[104] Fed-batch fermentation in 5 L and 1,500 L bioreactors produced 92.5 g/L and 81.2 g/L of L-arginine, respectively. The corresponding yields are 0.40 g and 0.35 g L-arginine per gram carbon source (glucose plus sucrose), respectively.[104] The lower production performance in 1,500 L bioreactor compared with that in 5 L bioreactor might partially due to the utilization of cheap industrial medium containing corn starch hydrolysate, raw sugar, and corn steep liquor.

3.8. Summary and Perspective

There is a large interest to develop bioprocesses due to the hazardous chemicals and substantial waste streams in the current chemical processes. The existing bioprocesses also need to be optimized and updated to satisfy the increasing demand and to be economically profitable. Based upon the technical platforms and knowledge accumulated in the past decades, the combination of traditional mutagenesis-screening methods and modern design-evolve-analysis strategies provide new promises for achieving these goals. Specifically, this could be achieved by integrating all the data gathered on the various dynamic interacting networks clarifying the biological systems.[105] Sustainable production of amino acids not only benefits the food/feed industrial, but also demonstrates the power of combining the fundamental research and industrial application.

References

1. Galli, F. Amino acid and protein modification by oxygen and nitrogen species. *Amino Acids* **32**, 497–499 (2007).
2. Brosnan, J.T. Amino acids, then and now — A reflection on sir Hans Krebs' contribution to nitrogen metabolism. *IUBMB Life* **52**, 265–270 (2001).
3. Suenaga, R. *et al.* Intracerebroventricular injection of L-arginine induces sedative and hypnotic effects under an acute stress in neonatal chicks. *Amino Acids* **35**, 139–146 (2008).
4. Wu, G. *et al.* Important roles for the arginine family of amino acids in swine nutrition and production. *Livest. Sci.* **112**, 8–22 (2007).
5. Westbrook, A.W., Moo-Young, M., Chou, C.P., & Kivisaar, M. Development of a CRISPR-Cas9 tool kit for comprehensive engineering of *Bacillus subtilis*. *Appl. Environ. Microbiol.* **82**, 4876–4895 (2016).
6. Araki, K., & Ozeki, T. *Kirk-Othmer Encyclopedia of Chemical Technology.* John Wiley & Sons, Inc., New York (2000).
7. Shimizu, H., & Hirasawa, T. *Amino Acid Biosynthesis Pathways, Regulation and Metabolic Engineering.* Springer, Heidelberg, 1–38 (2006).
8. Moran, E.T. Response of broiler strains differing in body fat to inadequate methionine: Live performance and processing yields. *Poult. Sci.* **73**, 1116–1126 (1994).

9. Murillo, M.G., & Jensen, L.S. Sulfur amino acid requirement and foot pad dermatitis in turkey poults. *Poult. Sci.* **55**, 554–562 (1976).

10. Sekiz, S.S., Scott, M.L., & Nesheim, M.C. The effect of methionine deficiency on body weight, food and energy utilization in the chick. *Poult. Sci.* **54**, 1184–1188 (1975).

11. Tabe, L., & Higgins, T. Engineering plant protein composition for improved nutrition. *Trends Plant Sci.* **3**, 282–286 (1998).

12. Kinoshita, S., Udaka, S., & Shimono, M. Studies on the amino acid fermentation. *J. Appl. Microbiol.* **3**, 193–205 (1957).

13. Leuchtenberger, W., Huthmacher, K., & Drauz, K. Biotechnological production of amino acids and derivatives: Current status and prospects. *Appl. Microbiol. Biotechnol.* **69**, 1–8 (2005).

14. Bongaerts, J., Krämer, M., Müller, U., Raeven, L., & Wubbolts, M. Metabolic engineering for microbial production of aromatic amino acids and derived compounds. *Metab. Eng.* **3**, 289–300 (2001).

15. Mavromichalis, I., Webel, D.M., Emmert, J.L., Moser, R.L., & Baker, D.H. Limiting order of amino acids in a low-protein corn-soybean meal-whey-based diet for nursery pigs. *J. Anim. Sci.* **76**, 2833–2837 (1998).

16. Ikeda, M. Economic aspects of amino acids production. In: R. Faurie *et al.* (eds.), *Microbial Production of L-Amino Acids*. Springer, Heidelberg, Berlin, pp. 1–35 (2003).

17. Sang, Y.L. Bio-based production of chemicals, fuels and materials by metabolically engineered microorganisms. *New Biotechnol.* **31**, S77 (2014).

18. Hermann, T. Industrial production of amino acids by coryneform bacteria. *J. Biotechnol.* **104**, 155–172 (2003).

19. Liu, H. *et al.* Effect of reducing protein level and adding amino acids on growth performance and carcass characteristics of finishing pigs. *J. Anim. Sci.* **77**, 69 (1999).

20. Kelle, R. *et al.* Glucose-controlled L-isoleucine fed-batch production with recombinant strains of *Corynebacterium glutamicum*. *J. Biotechnol.* **50**, 123–136 (1996).

21. Kumagai, H. *History of Modern Biotechnology*. Springer, New York, 71–85 (2000).

22. Eggeling, C. *et al.* Analysis of photobleaching in single-molecule multi-color excitation and Förster resonance energy transfer measurements. *J. Phys. Chem. A* **110**, 2979–2995 (2006).

23. Ikeda, M. Towards bacterial strains overproducing L-tryptophan and other aromatics by metabolic engineering. *Appl. Microbiol. Biotechnol.* **69**, 615 (2006).

24. Caldovic, L., & Tuchman, M. N-acetylglutamate and its changing role through evolution. *Biochemistry J.* **372**, 279–290 (2003).
25. Slocum, R.D. Genes, enzymes and regulation of arginine biosynthesis in plants. *Plant Physiol. Biochem.* **43**, 729–745 (2005).
26. de Jonge, W.J. *et al.* Arginine deficiency affects early B cell maturation and lymphoid organ development in transgenic ice. *J. Clin. Invest.* **110**, 1539–1548 (2002).
27. Ignarro, L.J., Buga, G.M., Wood, K.S., Byrns, R.E., & Chaudhuri, G. Endothelium-derived relaxing factor produced and released from artery and vein is nitric oxide. *Proc. Natl. Acad. Sci.* **84**, 9265–9269 (1987).
28. Traylor, T.G., & Sharma, V.S. Why nitric oxide? *Biochemistry* **31**, 2847–2849 (1992).
29. Kelle, R., Hermann, T., & Bathe, B. 20 L-Lysine production. In: L. Eggeling and M. Bott (eds.), *Handbook of Corynebacterium glutamicum.* CRC Press, Boca Raton, 465 (2005).
30. Gottschalk, G. *Bacterial Metabolism.* Springer-Verlag, New York, Xiii, 359 (1986).
31. Velasco, J., & Dobarganes, C. Oxidative stability of virgin olive oil. *Eur. J. Lipid Sci. Technol.* **104**, 661–676 (2002).
32. Eggeling, I., Cordes, C., Eggeling, L., & Sahm, H. Regulation of acetohydroxy acid synthase in *Corynebacterium glutamicum* during fermentation of α-ketobutyrate to L-isoleucine. *Appl. Microbiol. Biotechnol.* **25**, 346–351 (1987).
33. Lloyd-George, I., & Chang, T. Characterization of free and alginate-polylysine-alginate microencapsulated *Erwinia herbicola* for the conversion of ammonia, pyruvate, and phenol into L-tyrosine. *Biotechnol. Bioeng.* **48**, 706–714 (1995).
34. Ikeda, M., & Katsumata, R. Metabolic engineering to produce tyrosine or phenylalanine in a tryptophan-producing *Corynebacterium glutamicum* strain. *Appl. Environ. Microbiol.* **58**, 781–785 (1992).
35. Ikeda, M., & Katsumata, R. Hyperproduction of tryptophan by *Corynebacterium glutamicum* with the modified pentose phosphate pathway. *Appl. Environ. Microbiol.* **65**, 2497–2502 (1999).
36. Ito, H., Sato, K., Enei, H., & Hirose, Y. Improvement in microbial production of L-tyrosine by gene dosage effect of aroL gene encoding shikimate kinase. *Agric. Biol. Chem.* **54**, 823–824 (1990).
37. Chao, Y.P., Lai, Z.J., Chen, P., & Chern, J.T. Enhanced conversion rate of L-Phenylalanine by coupling reactions of aminotransferases and

phosphoenolpyruvate carboxykinase in *Escherichia coli* K-12. *Biotechnol. Prog.* **15**, 453–458 (1999).

38. Hummel, W., Schütte, H., Schmidt, E., Wandrey, C., & Kula, M.-R. Isolation of L-phenylalanine dehydrogenase from *Rhodococcus* sp. M4 and its application for the production of L-phenylalanine. *Appl. Microbiol. Biotechnol.* **26**, 409–416 (1987).

39. Nakamichi, K., Nabe, K., Nishida, Y., & Tosa, T. Production of L-phenylalanine from phenylpyruvate by Paracoccus denitrificans containing aminotransferase activity. *Appl. Microbiol. Biotechnol.* **30**, 243–246 (1989).

40. Choi, Y., & Tribe, D. Continuous production of phenylalanine using an *Escherichia coli* regulatory mutant. *Biotechnol. Lett.* **4**, 223–228 (1982).

41. Weikert, C., Sauer, U., & Bailey, J.E. Increased phenylalanine production by growing and nongrowing *Escherichia coli* strain CWML2. *Biotechnol. Prog.* **14**, 420–424 (1998).

42. Zeman, R. *et al.* Enzyme synthesis ofl-tryptophan. *Folia Microbiol.* **35**, 200–204 (1990).

43. Bang, W.G., Lang, S., Sahm, H., & Wagner, F. Production L-tryptophan by *Escherichia coli* cells. *Biotechnol. Bioeng.* **25**, 999–1011 (1983).

44. Faurie, R., & Fries, G. *Tryptophan, Serotonin, and Melatonin*. Springer, New York, 443–452 (1999).

45. Liebl, W. Corynebacterium taxonomy. *Handbook of Corynebacterium Glutamicum*. CRC Press, Boca Raton, FL, 9–34 (2005).

46. Liebl, W., Ehrmann, M., Ludwig, W., & Schleifer, K. Transfer of Brevibacterium divaricatum DSM 20297T, "Brevibacterium flavum" DSM 20411, "*Brevibacterium lactofermentum*" DSM 20412 and DSM 1412, and Corynebacterium lilium DSM 20137T to *Corynebacterium glutamicum* and their distinction by rRNA gene restriction patterns. *Int. J. Syst. Evol. Microbiol.* **41**, 255–260 (1991).

47. Udaka, S. Screening method for microorganisms accumulating metabolites and its use in the isolation of *Micrococcus glutamicus*. *J. Bacteriol.* **79**, 754 (1960).

48. Nakazawa, H., *et al.* Method of producing L-glutamic acid by fermentation. US patent 5,492,818 (1996).

49. Imaizumi, A., Kojima, H., & Matsui, K. The effect of intracellular ppGpp levels on glutamate and lysine overproduction in *Escherichia coli*. *J. Biotechnol.* **125**, 328–337 (2006).

50. Imaizumi, A. *et al.* Improved production of L-lysine by disruption of stationary phase-specific rmf gene in *Escherichia coli. J. Biotechnol.* **117**, 111–118 (2005).

51. Araki, K., & Nakayama, K. Studies on histidine fermentation: Part I. L-histidine production by histidine analog-resistant mutants from several bacteria. *Agric. Biol. Chem.* **35**, 2081–2088 (1971).

52. Kalinowski, J., Bachmann, B., Eggeling, L., Sahm, H., & Pühler, A. Genetic and biochemical analysis of the aspartokinase from *Corynebacterium glutamicum. Mol. Microbiol.* **5**, 1197–1204 (1991).

53. Thierbach, G., Kalinowski, J., Bachmann, B., & Pühler, A. Cloning of a DNA fragment from *Corynebacterium glutamicum* conferring aminoethyl cysteine resistance and feedback resistance to aspartokinase. *Appl. Microbiol. Biotechnol.* **32**, 443–448 (1990).

54. Tsuchida, T., & Momose, H. Genetic changes of regulatory mechanisms occurred in leucine and valine producing mutants derived from *Brevibacterium lactofermentum* 2256. *Agric. Biol. Chem.* **39**, 2193–2198 (1975).

55. Kisumi, M., Komatsubara, S., & Chibata, I. Valine accumulation by α-aminobutyric acid-resistant mutants of Serratia marcescens. *J. Bacteriol.* **106**, 493–499 (1971).

56. Tsuchida, T., & Momose, H. Improvement of an L-leucine-producing mutant of *Brevibacterium lactofermentum* 2256 by genetically desensitizing it to α-acetohydroxy acid synthetase. *Appl. Environ. Microbiol.* **51**, 1024–1027 (1986).

57. Chattopadhyay, M., Ghosh, A.K., & Sengupta, S. Control of methionine biosynthesis in *Escherichia coli* K12: A closer study with analogue-resistant mutants. *Microbiology* **137**, 685–691 (1991).

58. Kase, H., & Nakayama, K. L-Methionine production by methionine analog-resistant mutants of *Corynebacterium glutamicum. Agric. Biol. Chem.* **39**, 153–160 (1975).

59. Chattopadhyay, M., Ghosh, A.K., Sengupta, S., Sengupta, D., & Sengupta, S. Threonine analogue resistant mutants of *Escherichia coli* K-12. *Biotechnol. Lett.* **17**, 567–570 (1995).

60. Mondal, S., Das, Y., & Chatterjee, S. Methionine production by double auxotrophic mutants of a ethionine resistant strain of *Brevibacterium heali. Acta Biotech.* **14**, 61–66 (1994).

61. Kumar, D., & Gomes, J. Methionine production by fermentation. *Biotechnol. Adv.* **23**, 41–61 (2005).

62. Elišáková, V. *et al.* Feedback-resistant acetohydroxy acid synthase increases valine production in *Corynebacterium glutamicum*. *Appl. Environ. Microbiol.* **71**, 207–213 (2005).

63. Mendel, S. *et al.* Acetohydroxyacid synthase: A proposed structure for regulatory subunits supported by evidence from mutagenesis. *J. Mol. Biol.* **307**, 465–477 (2001).

64. Kopecký, J., Janata, J., Pospíšil, S., Felsberg, J., & Spížek, J. Mutations in two distinct regions of acetolactate synthase regulatory subunit from Streptomyces cinnamonensis result in the lack of sensitivity to end-product inhibition. *Biochem. Biophys. Res. Comm.* **266**, 162–166 (1999).

65. Park, J.H., Lee, K.H., Kim, T.Y., & Lee, S.Y. Metabolic engineering of *Escherichia coli* for the production of L-valine based on transcriptome analysis and in silico gene knockout simulation. *Proc. Natl. Acad. Sci.* **104**, 7797–7802 (2007).

66. Blombach, B. *et al.* L-Valine production with pyruvate dehydrogenase complex-deficient *Corynebacterium glutamicum*. *Appl. Environ. Microbiol.* **73**, 2079–2084 (2007).

67. Aiba, S., Tsunekawa, H., & Imanaka, T. New approach to tryptophan production by *Escherichia coli*: Genetic manipulation of composite plasmids *in vitro*. *Appl. Environ. Microbiol.* **43**, 289–297 (1982).

68. Azuma, S., Tsunekawa, H., Okabe, M., Okamoto, R., & Aiba, S. Hyper-production of L-trytophan via fermentation with crystallization. *Appl. Microbiol. Biotechnol.* **39**, 471–476 (1993).

69. Seiflein, T.A., & Lawrence, J.G. Methionine-to-cysteine recycling in Klebsiella aerogenes. *J. Bacteriol.* **183**, 336–346 (2001).

70. Sekowska, A., & Danchin, A. Identification of yrrU as the methylthioadenosine nucleosidase gene in *Bacillus subtilis*. *DNA Res.* **6**, 255–264 (1999).

71. Wheeler, P.R. *et al.* Functional demonstration of reverse transsulfuration in the *Mycobacterium tuberculosis* complex reveals that methionine is the preferred sulfur source for pathogenic mycobacteria. *J. Biol. Chem.* **280**, 8069–8078 (2005).

72. Nakamori, S., Kobayashi, S.-I., Kobayashi, C., & Takagi, H. Overproduction of L-cysteine and L-cystine by *Escherichia coli* strains with a genetically altered serine acetyltransferase. *Appl. Environ. Microbiol.* **64**, 1607–1611 (1998).

73. Utagawa, T. Production of arginine by fermentation. *J Nutr.* **134**, 2854S–2857S; discussion 2895S (2004).

74. Piérard, A., Glansdorff, N., & Yashphe, J. Mutations affecting uridine monophosphate pyrophosphorylase or the argR gene in *Escherichia coli*. *Mol. Genet. Genet. MGG* **118**, 235–245 (1972).

75. Gusyatiner, M.M., Leonova, T.V., Ptitsyn, L.R., & Yampolskaya, T.A. L-Arginine producing *Escherichia coil* and method of producing L-arginine. US Patent 7,052,884 (2006).

76. Hagishita, T., Yoshida, T., Izumi, Y., & Mitsunaga, T. Efficient L-serine production from methanol and glycine by resting cells of *Methylobacterium* sp. strain MN43. *Biosci. Biotechnol. Biochem.* **60**, 1604–1607 (1996).

77. Hashimoto, S.-I., & Katsumata, R. L-alanine fermentation by an alanine racemase-deficient mutant of the DL-alanine hyperproducing bacterium *Arthrobacter oxydans* HAP-1. *J. Ferment. Bioeng.* **86**, 385–390 (1998).

78. Katsumata, R., & Hashimoto, S.I. Process for producing alanine. US Patent 5,559,016 (1996).

79. Arita, M., Robert, M., & Tomita, M. All systems go: Launching cell simulation fueled by integrated experimental biology data. *Curr. Opin. Biotechnol.* **16**, 344–349 (2005).

80. Schilling, C.H. *et al.* Genome-scale metabolic model of *Helicobacter pylori* 26695. *J. Bacteriol.* **184**, 4582–4593 (2002).

81. Burgard, A.P., & Maranas, C.D. Probing the performance limits of the *Escherichia coli* metabolic network subject to gene additions or deletions. *Biotechnol. Bioeng.* **74**, 364–375 (2001).

82. Dennis, P.P., Ehrenberg, M., & Bremer, H. Control of rRNA synthesis in *Escherichia coli*: A systems biology approach. *Microbiol. Mol. Biol. Rev.* **68**, 639–668 (2004).

83. Edwards, J.S., Ibarra, R.U., & Palsson, B.O. In silico predictions of *Escherichia coli* metabolic capabilities are consistent with experimental data. *Nat. Biotechnol.* **19**, 125–130 (2001).

84. Ibarra, R.U., Edwards, J.S., & Palsson, B.O. *Escherichia coli* K-12 undergoes adaptive evolution to achieve in silico predicted optimal growth. *Nature* **420**, 186–189 (2002).

85. Lee, H.-S., & Hwang, B.-J. Methionine biosynthesis and its regulation in *Corynebacterium glutamicum*: Parallel pathways of transsulfuration and direct sulfhydrylation. *Appl. Microbiol. Biotechnol.* **62**, 459–467 (2003).

86. Nakamori, S., Kobayashi, S., Nishimura, T., & Takagi, H. Mechanism of L-methionine overproduction by *Escherichia coli*: The replacement of Ser-54 by Asn in the MetJ protein causes the derepression of L-methionine biosynthetic enzymes. *Appl. Microbiol. Biotechnol.* **52**, 179–185 (1999).

87. Rückert, C., Pühler, A., & Kalinowski, J. Genome-wide analysis of the L-methionine biosynthetic pathway in *Corynebacterium glutamicum* by targeted gene deletion and homologous complementation. *J. Biotechnol.* **104**, 213–228 (2003).

88. Leuchtenberger, W. Amino acids–technical production and use. *Biotechnology: Products of Primary Metabolism,* Vol. 6, 2nd Edition, Wiley-VCH Verlag GmbH, Weinheim, Germany, pp. 465–502 (2008).

89. Papin, J.A. *et al.* Comparison of network-based pathway analysis methods. *Trends Biotechnol.* **22**, 400–405 (2004).

90. Schilling, C.H., Letscher, D., & Palsson, B.Ø. Theory for the systemic definition of metabolic pathways and their use in interpreting metabolic function from a pathway-oriented perspective. *J. Theor. Biol.* **203**, 229–248 (2000).

91. Schuster, S., Dandekar, T., & Fell, D.A. Detection of elementary flux modes in biochemical networks: A promising tool for pathway analysis and metabolic engineering. *Trends Biotechnol.* **17**, 53–60 (1999).

92. Cox, S.J. *et al.* Development of a metabolic network design and optimization framework incorporating implementation constraints: A succinate production case study. *Metab. Eng.* **8**, 46–57 (2006).

93. Carlson, R., Fell, D., & Srienc, F. Metabolic pathway analysis of a recombinant yeast for rational strain development. *Biotechnol. Bioeng.* **79**, 121–134 (2002).

94. Duarte, N.C., Palsson, B.Ø., & Fu, P. Integrated analysis of metabolic phenotypes in *Saccharomyces cerevisiae. BMC Genom.* **5**, 63 (2004).

95. Liao, J.C., & Oh, M.-K. Toward predicting metabolic fluxes in metabolically engineered strains. *Metab. Eng.* **1**, 214–223 (1999).

96. Carlson, R., & Srienc, F. Fundamental *Escherichia coli* biochemical pathways for biomass and energy production: Creation of overall flux states. *Biotechnol. Bioeng.* **86**, 149–162 (2004).

97. Ibarra, R., Fu, P., Palsson, B., DiTonno, J., & Edwards, J. Quantitative analysis of *Escherichia coli* metabolic phenotypes within the context of phenotypic phase planes. *J. Mol. Microbiol. Biotechnol.* **6**, 101–108 (2004).

98. Vijayasankaran, N., Carlson, R., & Srienc, F. Metabolic pathway structures for recombinant protein synthesis in *Escherichia coli. Appl. Microbiol. Biotechnol.* **68**, 737 (2005).

99. Park, J.H., & Lee, S.Y. Fermentative production of branched chain amino acids: A focus on metabolic engineering. *Appl. Microbiol. Biotechnol.* **85**, 491–506 (2010).

100. Park, J.H., Kim, T.Y., Lee, K.H., & Lee, S.Y. Fed-batch culture of *Escherichia coli* for L-valine production based on in silico flux response analysis. *Biotechnol. Bioeng.* **108**, 934–946 (2011).

101. Rieping, M., & Hermann, T. L-Threonine. In: *Amino Acid Biosynthesis Pathways, Regulation and Metabolic Engineering.* Springer Berlin Heidelberg, 71–92 (2006).

102. Hermann, T. Industrial production of amino acids by coryneform bacteria. *Journal of biotechnology* **104**, 155–172 (2003).

103. Ohnishi, J., Hayashi, M., Mitsuhashi, S., & Ikeda, M. Efficient 40 degrees C fermentation of L-lysine by a new *Corynebacterium glutamicum* mutant developed by genome breeding. *Appl. Microbiol. Biotechnol.* **62**, 69–75 (2003).

104. Park, S.H. *et al.* Metabolic engineering of Corynebacterium glutamicum for L-arginine production. *Nat. Commun.* **5**, 4618 (2014).

105. Vertès, A.A., Inui, M., & Yukawa, H. Manipulating corynebacteria, from individual genes to chromosomes. *Appl. Environ. Microbiol.* **71**, 7633–7642 (2005).

Chapter 4

Fatty Acid Biosynthesis and Its Metabolic Engineering Applications

Yi Liu* and Tiangang Liu*,†,‡

**Key Laboratory of Combinatorial Biosynthesis
and Drug Discovery, Ministry of Education
and Wuhan University School of Pharmaceutical Sciences
Wuhan, China
†Hubei Engineering Laboratory for Synthetic Microbiology
Wuhan Institute of Biotechnology, Wuhan, China
‡liutg@whu.edu.cn*

4.1. Introduction

Fatty acids (FAs) are important metabolites derived from highly complex and energy-intensive lipid metabolism in living cells. With extensive investigation of mechanisms associated with FA biosynthesis (FAB), the bioproduction of FA-derived chemicals is of particular interest for metabolic engineering applications. To convert FAs into an array of chemicals with useful properties, rapidly developed synthetic biology techniques are employed to enable the capture of suitable microbes to engineer metabolic pathways. FAs could be further reduced to fatty alcohols (FALs), esterified to esters, and decarboxylated to alkanes or alkenes to yield renewable alternatives to petrochemicals.[1] Also, increasing attention has been paid to

97

the conversion of FAs to other chemicals, such as biomaterials [polyhydroxyalkanoates (PHA)], nutritional supplements, and drug precursors. The current trend toward commercialization is expected to achieve wide future applications in the fields of industry, medicine, and agriculture. Currently, new perspectives and new findings from omics-based studies are facilitating our understanding of the complex network of FA metabolism, leading to remarkable progress in yield and productivity. This chapter will highlight FA metabolism, as well as how scientists approach the challenges of microbial production of valuable FA-derived chemicals, as well as identify the associated bottlenecks.

4.2. Fundamentals of FAB in Biological Systems

FAs are carboxylic acids with long aliphatic tails (chains) and are known as integral components of living systems. They are implicated in a variety of important intracellular metabolic activities, such as membrane synthesis, energy storage, post-translational protein modifications, and serve as secondary metabolites and signaling molecules. Several FAB-related enzymes have been identified as reasonable drug targets during development of new antibacterials.[2,3] Due to its importance, FAB is one of the most ubiquitous and highly conserved pathways in organisms. In recent decades, FAB systems in microbes have received significant attention, and the metabolic mechanism has been extensively characterized, especially in the commonly used strain *Escherichia coli*.

4.2.1. *FA Metabolism*

De novo synthesis of FAs is catalyzed by FA synthases (FASs). Based on their architecture, FASs can be divided into two classes: types I and II, which are mainly present in eukaryotes and prokaryotes, respectively. Type I FAS systems exist in animals, fungi, and some bacteria, such as *Mycobacteria* and *Corynebacteria*,[4] whereas type II FASs are found mostly in bacteria, but also in eukaryotic organelles, such as mitochondria and plastids. In some actinomycetes, such as *Mycobacterium* species, both types I and II FASs are present.[5] The type I system is characterized by the use of a large multifunctional protein complex that carries all proteins necessary for FAB on one or two large polypeptide chains. The crystal structure of *Saccharomyces cerevisiae* FAS, a member of the fungal type I FAS family, has been determined.[6] In the case of the type II system, the individual enzymatic reaction is carried out by a set of mono-functional proteins encoded by unique genes. The high-resolution X-ray and/or nuclear magnetic resonance structures of every enzyme in the *E. coli* type II FAS pathway have been thoroughly analyzed.[7] These structural studies led to a deep understanding of the reaction mechanism. Although the organization of the FAS system varies between different organisms, the profiles of FAB are essentially the same, with FAS types I and II FAS sharing common mechanisms during the initiation and elongation stages. The profiles of FAB in *E. coli* and *S. cerevisiae* are shown in Figures 4.1 and 4.2, respectively.

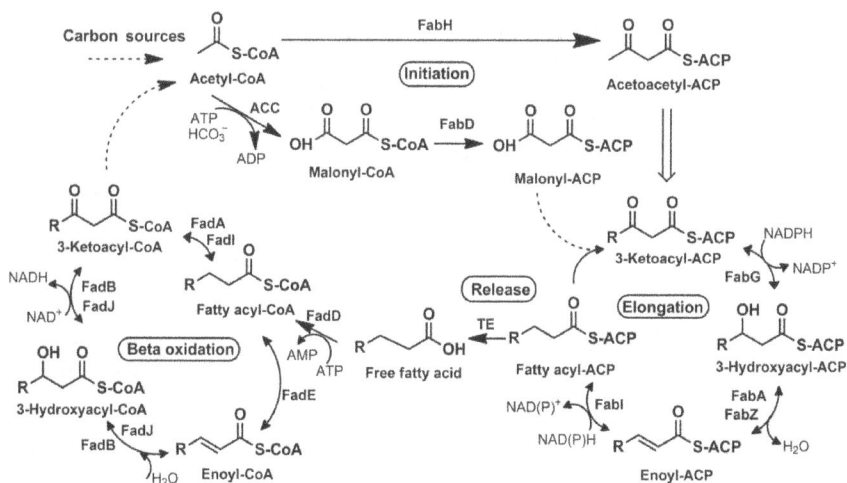

Figure 4.1.   FAB in *E. coli*. ACC, acetyl-CoA carboxylase; FabD, malonyl-CoA, ACP, transacylase; FabH, 3-ketoacyl-ACP synthase III; FabG, 3-ketoacyl-ACP reductase; FabB, 3-ketoacyl-ACP synthase I; FabZ/FadA, 3-hydroxyacyl-ACP dehydratase; FabI, enoyl-acyl-ACP reductase; FadD, acyl-CoA synthetase; TE, thioesterase; FadE, acyl-CoA dehydrogenase; FadJ/FadB, enoyl-CoA hydratase/3-hydroxyacyl-CoA dehydrogenase; FadA/FadI, acetyl-CoA-acyltransferase.

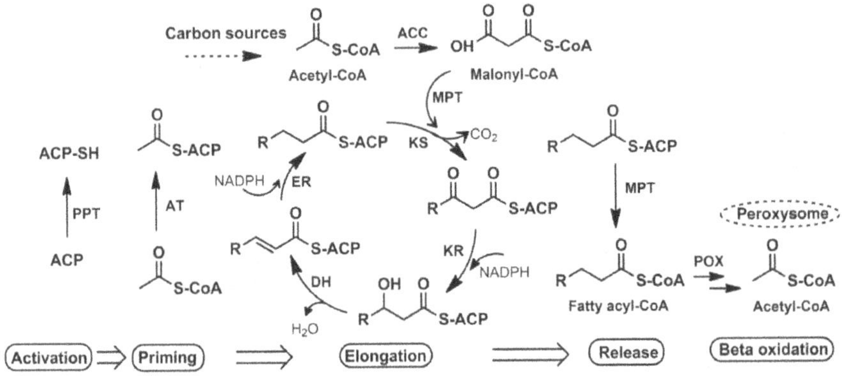

Figure 4.2. FAB in *S. cerevisiae*. ACC, acetyl-CoA carboxylase; PPT, phosphopantetheinyl transferase; AT, acetyltransferase; KS, ketoacyl synthase; MPT, malonyl/palmitoyl transferase; KR, ketoacyl reductase; DH, dehydratase; ER, enoyl reductase; POX, fatty-acyl coenzyme A oxidase.

4.2.1.1. *FAB in E. coli*

The canonical *E. coli* FAB pathway begins with the conversion of acetyl-CoA into malonyl-CoA by acetyl-CoA carboxylase (ACC) at the cost of adenosine triphosphate (ATP), which is one of the rate-limiting steps in FAB.[8] Malonyl-CoA is then transferred to malonyl-acyl carrier protein (ACP) by malonyl-CoA:ACP transacylase (FadD). The generated malonyl-ACP is used as an extender unit for the elongation stage. The most acceptable mechanism for the initiation of FAB is FabH (3-ketoacyl-ACP synthase III), which catalyzes the condensation of acetyl-CoA with malonyl-ACP to yield acetoacetyl-ACP, followed by reduction to 3-hydroxyacyl-ACP by a nicotinamide adenine dinucleotide (NADPH)-dependent 3-ketoacyl-ACP reductase (FabG). In the elongation cycle, the first step is the Claisen condensation of malonyl-ACP with fatty acyl-ACP, which is catalyzed by a 3-ketoacyl-ACP synthase (either FabB or FabF) to form 3-ketoacyl-ACP. This is the only irreversible step and, therefore, 3-ketoacyl-ACP synthase is the key point for regulating the product distribution of the pathway. 3-Ketoacyl-ACP is subsequently reduced by FabG to produce 3-hydroxyacyl-ACP and then further reduced to enoyl-ACP by a 3-hydroxyacyl-ACP dehydratase (FabZ or FabA). The final reduction is catalyzed by an enoyl-ACP reductase (FabI), resulting in acyl-ACP. Next, another round of elongation begins, using acyl-ACP as a substrate. All these reduction steps require NADPH as a cofactor. It is notable that all of the pathway intermediates are shuttled between the enzymes while bound to the small, acidic, and extremely soluble ACP. The FAB pathway stops when a certain chain length (C14 to C18, but typically C16) is reached. Afterward, the acyl chains can be hydrolyzed to free FAs (FFAs) by the TE TesA. Some released FFAs are activated by acyl-CoA synthetase (FadD) to form fatty acyl-CoA, which can be trans-formed to complex lipids, such as phospholipids and lipid A, or enter the β-oxidation cycle. During this cyclical reaction, an acyl-CoA dehydroge-nase (FadE) catalyzes fatty acyl-CoA to form enoyl-CoA, which is hydrated and dehydrogenated by an enoyl-CoA-hydratase (FadB or FadJ) to 3-hydroxyacyl-CoA and 3-ketoacyl-CoA. Finally, fatty acyl-CoA is derived from 3-ketoacyl-CoA thiolase (FadA or FadI) catalysis to form a new acyl-CoA containing two fewer carbon atoms than the original molecule.[9]

In general, the FA metabolic process includes initiation, elongation, release, and an additional degradation pathway (β-oxidation). In *E. coli*, the cycle continues in an iterative manner, operating on ACP. During each iteration, two carbons are added from malonyl-ACP to a growing acyl-chain, and the resulting 3-keto group is reduced to a saturated methylene.

The regulation of FA metabolism is tightly controlled at multiple levels during cell growth for proper function under various conditions. The primary regulation of FA metabolic pathways is conducted at the transcriptional level. The key unsaturated FAB genes, *fabA*, and *fabB*, are controlled by FadR and FabR, the two transcriptional regulators of type II FAS in *E. coli*.[10,11] Additionally, FAS-related enzymes ACC, FabH, and FabI are feedback inhibited by long-chain fatty acyl-ACPs.[11,12] Overall, these mechanisms ensure that the cell does not accumulate excess quanti-ties of this energy-rich hydrocarbon.

4.2.1.2. *FAB in S. cerevisiae*

There are several differences between type I and type II FASs in *de novo* FAB. In the case of *S. cerevisiae*, primarily C16 and C18 FAs are produced by the large 2.6 MDa type I cytosolic FAS. Six copies of α and β subunits (FAS2 and FAS1) organize as a barrel-shaped $\alpha6\beta6$ complex containing eight independent functional domains. These functional domains catalyze all reactions required for the biosynthesis of FAs in yeast: activation, priming, multiple cycles of elongation, and termination. The coordinated synthesis occurs inside the $\alpha6\beta6$ complex, leading to kinetically more efficient and readily controlled catalytic reactions relative to those of dissociated type II FASs.[4] Briefly, acetyl-CoA generated through the direct activation of acetate in the cytosol is converted into malonyl-CoA by ACC. With malonyl-CoA as a building block, fatty acyl-CoA is gradually extended two carbons at a time in the FAS complex until a chain length of C16–C18 is attained. First, FA synthesis is activated by a phosphopantetheinyl transferase domain catalyzing the activation of ACP. Subsequently, the acetyltransferase domain catalyzes the priming reaction by loading the acetyl moiety from acetyl-CoA onto the FAS-reactive thiol group of the ACP pantetheine arm, followed by its being shuttled to the catalytic site of the ketoacyl synthase (KS) domain for chain elongation. After the transfer of a malonyl moiety from malonyl-CoA by malonyl/palmitoyl transferase (MPT), KS condenses its bound acetyl-starter group with the ACP-attached malonyl portion to 3-ketoacyl-ACP. Next, the ketoacyl reductase domain reduces it to a 3-hydroxyl intermediate, followed by a dehydration reaction catalyzed by the dehydratase (DH) domain to yield a 3-enoyl moiety. A second reduction reaction is catalyzed by the enoyl reductase domain to yield a saturated acyl chain elongated by a two-carbon unit. ACP shuttles the reaction intermediates and brings this acyl product back to the KS domain for use as a primer substrate with another malonyl-ACP for the next round of elongation. Termination occurs when the length of the FA chain reaches 16 or 18 carbons, and the bifunctional MPT back-transfers the products to coenzyme A for release.[13] Notably, the final released products are fatty acyl-CoAs that are distinct from the fatty acyl-ACPs in *E. coli*. Additionally, fatty acyl-CoAs are transported intracellularly to the endoplasmic reticulum as lipid bodies for phospholipid or triacylglycerol biosynthesis. Additionally, β-oxidation of FAs mainly occurs in peroxisomes using the activated precursor acyl-CoA.[14] To ensure proper FA composition and homeostasis, regulation of these processes is achieved by regulating the expression of essential genes, such as *Acc1*, *Fas1*, and *Fas2*, at multiple levels.[15]

4.2.2. *FA Derivatives*

Although FAs are primarily implicated in phospholipid biosynthesis and maintaining organism growth, a wide range of FA derivatives can meet our requirements in product uses. These FAs can be applied as biofuels, but also as ingredients in food, cosmetics, or for the pharmaceutical industry. Due to increasing concerns related to global environmental sustainability, large commercial applications of FA-derived products using microbial FASs were investigated in recent years.[1,16,17] FAs or their activated forms, fatty acyl-CoAs, and fatty acyl-ACPs, are important precursors that allow synthesis of a variety of next-generation biofuels and chemicals, including FFAs, FALs, FA methyl/ethyl esters (FAMEs/FAEEs), and alkanes. Different examples will be presented in the following sections.

4.3. Applications of FA Metabolic Engineering for Chemical Synthesis in Microbes

FA synthesis occurs according to complex biosynthetic machinery, with strict regulation at multiple levels. Also, substrates, intermediates, and products often cause chemical toxicity, interfere with metabolic processes, and damage cellular infrastructure. Therefore, the production of high-value compounds requires accurate design of engineering strategies to improve microorganism viability and the optimization of metabolic pathways to maximize productivity and yield. Furthermore, for optimal performance, it is essential to have a balanced expression of metabolic pathways. Scientists responded to these challenges in creative ways.

The concept of a microbial cell factory, engineered microbes, or microbial consortia is alive in the field of metabolic engineering. *E. coli*, cyanobacteria, *S. cerevisiae*, and oleaginous yeasts have been explored as popular cell factories. Microbial cells can produce not only conventional metabolites, such as proteins, fuels, drugs, and food, but also biomaterials with important applications in emerging biomedical fields. Apart from

E. coli as the most popular engineered strain, *S. cerevisiae* has been successfully used in modern fermentation industry thanks to its well-studied genetic and physiological background, the availability of a large collection of genetic tools, the compatibility of high-density and large-scale fermentation, resistance to phage infection, and high tolerance against toxic inhibitors and products. Currently, increasing efforts focus on yeast.

In current research associated with microbial FA metabolic engineering, conversion of renewable feedstock into fuels and chemicals from microbial cell factories focuses on construction of metabolic pathways and balance of the entire metabolic network for target compounds. These methods are based on fundamental investigations, including functional verification of the key enzymes and identification of the key players in the metabolic pathway. With developments in over more than 20 years, the strategies and technologies associated with metabolic engineering are well established and initial success has been achieved.

4.3.1. *In Vitro Analysis of FA Synthesis*

Advances in traditional engineering have been achieved in recent years; however, increased study led to increased difficulties. Due to the lack of steady-state kinetic information and incomplete understanding of the biosynthetic pathway, it remains difficult to build a highly efficient and universal system.

In 2010, Liu *et al.* developed a cell-free system that could be used for direct quantitative investigation of FAB and regulation in *E. coli*. It is a more reliable engineering strategy that avoids the *in vivo* metabolic barrier.[8] Cell-free extracts from *E. coli* were prepared for *in vitro* assays, and by titrating additional concentrations of substrates, intermediates, cofactors, and enzymes into the *in vitro* cell-free system, potential metabolic bottlenecks were examined. Specifically, it was discovered that malonyl-CoA levels, rather than NADPH levels, were limiting. Generally, cell-free systems could also be used to identify bottlenecks in FAB in other organisms, such as algae, plants, and yeast.[18] In 2011, Yu and colleagues determined the limits of the pathway for FA production by *in vitro* reconstitution of *E. coli* FAS. Purified protein components of *E. coli* FAS were used for detailed kinetic analysis in this reconstituted system, enabling individual enzymes in the FAB pathway to be examined and their roles studied by *in vitro* steady-state analysis.[19] That work highlighted the utility of a cell-free system for investigating the properties of FA synthases under steady-state conditions.

These typical *in vitro* studies provided fundamental knowledge for engineering FAB and indicated that by altering the ratio of FAS subunits, it was possible to increase FA yields.

4.3.2. *Overproduction of FFAs*

To meet the commercialized demand of FA-derived fuels and chemicals, the first concern involves engineering of metabolic pathways to maximize the production of FFAs. Several groups exerted significant effort to increase the metabolic flux of FA synthesis and improve the accumulation of FFAs based on the knowledge of FA regulation.

An encouraging work concerning overproduction of FA in *E. coli* was first reported by Lu and colleagues in 2008.[20] In their strategy, four distinct genetic alterations were systematically introduced into a single host, including elimination of β-oxidation by knocking out the endogenous gene *fadD*, heterologous expression of a plant thioesterase (TE) to increase the abundance of shorter-chain FA, overexpression of *E. coli* ACC to increase the supply of malonyl-CoA, and overexpression of an endogenous TE (TesA), a "leaderless" TesA variant, to release feedback inhibition caused by long chain acyl-ACPs. Finally, 2.5 g/L FAs were produced, with a productivity rate of 0.024 g/h/g dry cell mass under fed-batch fermentation conditions. All these modifications yielded a 19-fold increase in total FA production as compared to that attained using the original strain. At least 50% of the FAs produced were present in FFA form in this engineered *E. coli* strain. This result from *E. coli* opened the door to harnessing the power of molecular genetics and bioprocess engineering in experimentally friendly organisms. Scientists continued searching for more efficient methods to improve the system.

Previous studies demonstrated the dynamic regulation of the FAB pathway. FadR is a key transcription factor that controls the expression of several FAB genes. Zhang *et al.* applied a global view of FAB and focused on regulatory proteins that affected FAB and FA degradation.[21] They achieved significant advances in enhancing FA production through the expression of the regulatory transcription factor FadR. By tuning the expression of FadR in an engineered *E. coli* host, FA production increased to 5.27 g/L, or 73% of the theoretically possible yield. Whole-genome transcriptional analysis using microarrays and targeted proteomics was performed to elucidate the mechanism by which FadR enhanced FA yield. Overexpression of FadR led to transcriptional changes in genes involved in FA pathways, especially *fabB*, *fabF*, and *accA*; however, overexpression of any of these genes alone did not achieve the high yields resulting from *fadR* expression, indicating that FadR enhanced FA production globally by regulating the expression of many genes to optimal levels. This method to improve FA yield by overexpressing FadR can be used to increase the cellular FA pool, thereby providing additional precursors for the production of FA-derived biofuels.

Moreover, with the perspective of global control, researchers used a modular engineering approach instead of enhancing or knocking down expression of individual genes.[22] Based on central pathway architecture, *E. coli* FAB was deconstructed into three modules: the upstream acetyl coenzyme A-formation module (GLY module), the intermediary acetyl-CoA-activation module (ACA module), and the downstream FAB module (FAS module). An optimized combination of these three modules was successfully expressed with varying promoter strengths and plasmid-copy numbers. Subsequently, refining protein translation efficiency by customizing ribosome-binding sites for both the upstream GLY and FAS modules enabled further production improvement. After optimizing cultivation conditions in a fed-batch 20-L fermenter, the engineered strain produced 8.6 g/L of FFAs. The modular engineering strategies systematically removed metabolic pathway bottlenecks and led to significant titer improvements in a multigene FA metabolic pathway, demonstrating a generalized approach to engineering cell factories for valuable metabolite production.

In addition to the above methods, an innovative approach was established to synthesize biofuels by reversing β-oxidation through engineering the *E. coli* β-oxidation cycle as a metabolic platform for the synthesis of advanced fuels and chemicals.[23] By combining endogenous dehydrogenases and TEs, this pathway bypassed the biosynthesis of malonyl-CoA and was employed to produce FAs at high yields (7 g/L).

To date, most of the work has taken a static perspective, including modification of plasmid copy number,[24] promoter strength,[25] and combinations of these strategies,[26] to coordinate the expression of enzymes and optimize production titer and yield. A novel approach was designed using a dynamic regulatory network in *E. coli* to switch the supply and consumption of malonyl-CoA and efficiently redirect carbon flux toward FAB. This control resulted in an oscillatory malonyl-CoA pattern and a balanced metabolism between cell growth and product formation.[27] *S. cerevisiae* is also an important industrial host for the production of enzymes, pharmaceutical and nutraceutical ingredients, and commodity chemicals.[28] *S. cerevisiae* has many advantages for industrial production as compared to *E. coli* and has already been engineered to produce biofuels.[29–31] In 2014, Li *et al.* engineered *S. cerevisiae* to overproduce FAs, resulting in 140 mg/L FAs produced by the engineered strain in flask-shake cultures.[18] Additionally, ACC was proven to be a rate-limiting step in FA synthesis in a yeast cell-free system. These efforts were crucial for developing a "cell factory" for the overproduction of FAs using type I FAS. In the chemical industry, short branched-chain FAs (SBCFAs, C4–C6) are versatile platform intermediates enabling the production of value-added products and are mainly synthesized chemically, which is costly and damages the environment. In order to develop green production, Yu *et al.* engineered *S. cerevisiae* to overproduce SBCFAs via a combinatorial metabolic engineering approach by optimizing the native Ehrlich pathway. Through incorporation of a chromosome-based combinatorial gene-overexpression strategy, elimination of genes in competitive pathways, and overexpression of a native transporter, the production of SBCFAs was increased to 387.4 mg/L.[32] Currently, new yeast cell factories have been developed for the production of FFAs and the subsequent production of alkanes and FALs from its descendants.[33] The engineered strains can produce up to 10.4 g/L of FFAs, which is the highest production titer reported to date. These examples provide evidence that FA production can be increased step by step in a model organism through different metabolic engineering strategies.

4.3.3. *Microbial Production of FA-Derived Fuels and Chemicals*

Microorganisms have been engineered to convert simple sugars into several types of biofuels and chemicals, such as alcohols, FA esters, and alkanes (Figure 4.3). Here, we reviewed recent efforts on the production of these advanced biofuels in the well-characterized microorganisms *E. coli* and *S. cerevisiae,* resulting in high titers and yields.

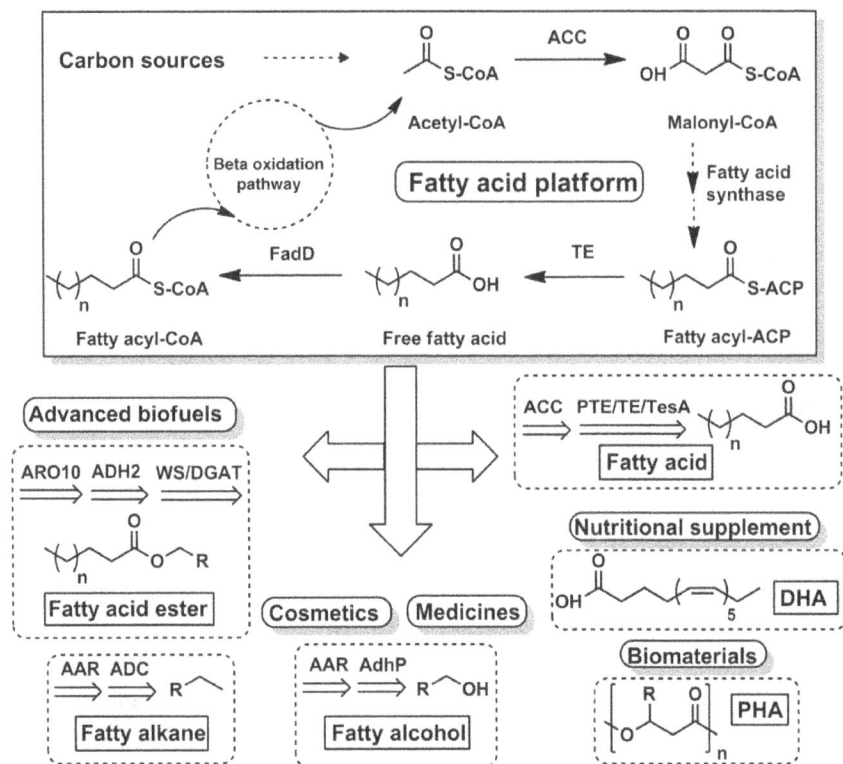

Figure 4.3. FA platform for downstream products. ACC, acetyl-CoA carboxylase; TE, thioesterase; FadD, acyl-CoA synthetase; ADC, aldehyde decarbonylase; AAR, acyl-ACP reductase; ARO10, 2-keto acid decarboxylase; ADH2, alcohol dehydrogenase; AdhP, ethanoldehydrogenase/alcoholdehydrogenase; WS/DGAT, wax ester synthase/acyl-CoA, diacylglycerol acyltransferase; PTE, TE, and TesA, thioesterases from *S. cerevisiae, Cinnamomum camphorum* and *E. coli.*

4.3.3.1. *FALs*

FALs and their derivatives have been commercially applied in the fields of chemistry and the chemical industry. They can be used as surfactants, lubricants, plasticizers, solvents, emulsifiers, flavors, and fragrances, as well as for the formulation of pharmaceuticals.[34,35] FALs were traditionally produced by direct extraction from plant material or chemical synthesis from fossil sources. In contrast, current microbial production without competition with food oil production enables tailored production of specific FALs.

As non-fossil-fuel energy sources, long-chain FALs possess high energy density relative to shorter-chain biofuel ethanol. FALs can be reduced from fatty acyl-ACPs, fatty acyl-CoAs, or FAs by the enzyme fatty acyl-CoA/ACP reductase[36] or carboxylic acid reductase,[37] with fatty aldehydes as the intermediates that can be further catalytically converted to hydrocarbons, such as alkenes.[38] In 2013, a *Mycobacterium marinum* carboxylic acid reductase (CAR) was found to convert a wide range of aliphatic FAs (C6–C18) into corresponding aldehydes. By expressing CAR and an aldehyde reductase in an engineered *E. coli* strain, an FAL titer exceeding 350 mg/L was obtained.[37] In 2014, Liu *et al.* engineered a fatty acyl-ACP reductase-dependent pathway to improve FAL production in *E. coli*.[39] Fatty acyl-CoA/ACP reductase is a key enzyme found in many organisms.[40] It was hypothesized that the use of fatty acyl-ACP as a substrate would be the most economical strategy for ATP consumption; therefore, the *in vitro* system was reconstituted. Based on substrate preferences, four fatty acyl-CoA/ACPs reductases, *Simmondsia chinensis* (jojoba) FAR,[41] *Acinetobacter calcoaceticus* Acr1,[42] *Oryza sativa* DPW,[43] and *Synechococcus elongatus*acyl-ACP reductase (AAR)[38] were selected for *in vitro* reconstitution of the FAL biosynthetic pathway. As a result of metabolic engineering strategies, an *E. coli* mutant containing AAR improved FAL yields and productivity rates to 750 mg/L and 0.06 g/L/h, respectively. Both *in vitro* and *in vivo* results demonstrated that the activity and expression level of fatty acyl-CoA/ACP reductase was the rate-limiting step in the current protocol. This case established a promising *in vitro* reconstitution-based synthetic pathway for industrial microbial production of FALs.

4.3.3.2. *Fatty alkanes/alkenes*

Alkanes/alkenes with different chain lengths are valued as the main components of diesel fuels naturally produced by diverse organisms. For example, in higher plants, alkanes are mainly involved in synthesizing the epicuticula wax layer to reduce water loss.[44] Alkanes/alkenes with C4–C23 carbon-chain length possess higher energy densities, hydrophobic properties, and are compatible with existing liquid-fuel infrastructure, including gasoline, petrodiesel, and jet fuels, used in vehicles.

In 2010, two key enzymes, which together convert intermediates of FA metabolism to alkanes and alkenes implicated in fatty alkane/alkene formation, were identified in cyanobacteria. Acyl-ACP can be reduced to fatty aldehyde by an AAR, followed by aldehyde oxidation to alkane or alkene by an aldehyde decarbonylase. Heterologous expression of these two genes in an engineered *E. coli* strain resulted in production of a mixture of alkanes and alkenes (C13–C17), with a yield of 300 mg/L.[38] Thus far, there have been several reported pathways for engineering FA-derived alkane and alkene production,[38,45–49] as well as a reversed β-oxidation pathway demonstrated by an *in silico*-modeling study.[50] By engineering *E. coli* FAB and degradation pathways and employing an engineered TE, short-chain alkanes can be produced at levels of up to 580.8 mg/L through fatty acyl to FA to fatty acyl-CoA pathways.[46] After optimizing the FFA-derived pathway catalyzed by a cytochrome P450 OleTJE, the total hydrocarbon titer obtained was 97.6 mg/L.[51] However, these products are in mixed forms due to the nature of the type II FAB mechanism. In 2015, Liu *et al.* engineered a pathway for the iterative type I polyketide synthase SgcE with the cognate TE, SgcE10, in *E. coli*, resulting in a pentadecan yield of nearly 140 mg/L in single-alkane form through fed-batch fermentation.[52]

4.3.3.3. *FA esters*

FAEE is an excellent diesel fuel replacement due to its low water solubility and high energy density and is suitable for microbial production due to its low toxicity to host cells.[53]

The initial identification of a bifunctional wax ester synthase/acyl-CoA:diacylglycerol acyltransferase (WS/DGAT) catalyzing the biosynthesis of wax esters and triacylglycerols triggered a series of engineering works to synthesize FAEEs from alcohols and fatty acyl-CoAs.[54,55] In *E. coli*, 0.26 g/L FAEEs was achieved using exogenous oleic acid.[56] Later, 674 mg/L FAEEs was obtained from plant-derived biomass, such as hemicellulose, instead of feeding Fas.[57] Additionally, fed-batch cultivation and a dynamic sensor-regulator system improved FAEE production to 0.922 and 1.5 g/L, respectively.[58,59] Moreover, some investigations focused on the production of medium-chain esters, such as isobutyl acetate and isoamyl acetate, as well as butyrate esters, in *E. coli*.[60,61] Furthermore, *S. cerevisiae* was also engineered for FAEE production by condensation of fatty acyl-CoAs and ethanol. After metabolic engineering with the modification of culture conditions, a maximum titer of >25 mg/L FAEEs was produced.[62] Nielsen's group[63] successfully expressed five WSs derived from different organisms in *S. cerevisiae* separately and evaluated the potential for FAEE production. Expression of WS from *Marinobacter hydrocarbonoclasticus* DSM8798 resulted in a high titer of 6.3 mg/L. Next, they achieved FAEE production of 34 mg/L by chromosome engineering, as well as an additional 40% increase in FAEE production.[64] Another work from Atsumi's group reported the biosynthesis of a multitude of esters by combining various alcohol-related biosynthetic pathways and a diversity of CoA molecules, achieving high production of isobutyl acetate from glucose (17.2 g/L).[65]

Also, FA short-chain esters (FASEs) are renewable, nontoxic, and biodegradable biofuels. Recently, a novel approach for the biosynthesis of FASEs was developed using an engineered *E. coli* strain in combination with the FA and 2-keto acid pathways. FASE biosynthesis in *E. coli* was achieved by expression of WS/DGAT, which catalyzes the esterification of fatty acyl-CoAs and short-chain alcohols. Short-chain alcohols were synthesized from the 2-keto acid pathway by expression of 2-keto acid decarboxylase and an alcohol dehydrogenase. After additional engineering, the optimal strain produced a titer of 1,008 mg/L FASEs in fed-batch cultivation. Because the FA and 2-keto acid pathways are native in microbes, this strategy will apply to other microorganisms to produce various FASEs from renewable raw materials, such as sugars and cellulose.[66] Due to their improved low-temperature properties, *de novo* biosynthesis of FA branched-chain esters (FABCEs) and branched FABCEs was performed in an engineered *E. coli* strain through combination of the branched FAB pathway and the branched-chain amino acid biosynthetic pathway. By modifying the FA pathway, FABCE production was improved to 273 mg/L and achieved a high proportion of FABCEs at 99.3% of total FAEEs. Additionally, *Pichia pastoris* yeast was also engineered to produce 169 mg/L FABCEs for the first time.[67]

4.3.3.4. *PHA*

PHAs are microbially synthesized polyesters, which can be used as biodegradable plastics and carbon-storage materials. Recently, metabolic engineering studies of microbial pathways associated with PHA production were performed. A metabolic pathway was reported for converting glucose into medium-chain-length PHA, which combined FAB, an acyl-ACP TE to generate the desired C12 and C14 FAs, β-oxidation for conversion of FAs to (R)-3-hydroxyacyl-CoAs, and a PHA polymerase. The yields improved through the expression of an acyl-CoA synthetase, resulting in production of more than 15% cell dry weight.[68] Another *E. coli* PHA-production system was developed that exhibited control over repeating-unit composition for both medium-chain-length and short-chain-length PHAs.[69] The challenge to commercial production of these renewable, biodegradable PHAs is using inexpensive feedstocks by engineering the pathways to allow competitive, cost-effective production of petroleum-derived plastics.

4.4. Perspectives and Prospects of the FA Platform

Microbes are intrinsically capable of synthesizing FAs, which can be used as precursors for a number of industrially relevant molecule classes, such as FALs, alkanes, and different types of esters. These can be applied as biofuels, but also as ingredients in food, cosmetics, and the pharmaceutical industry. This chapter cited some representative cutting-edge research associated with comprehensive efforts to elucidate metabolic engineering approaches to optimize FA conversion to biofuels. Although facing immense social and environmental problems, there is great potential here, as well.

Until now, metabolic engineering strategies have led to certain achievements in production of high-value FA-derived products. FA-based systems have been well established and engineered for future fuel feedstock in model organisms. For example, sustainable production of FA-derived fuels and chemicals requires establishment of cell-factory platform strains. Currently, new results showed the highest FFA titer (10.4 g/L) in *S. cerevisiae* by systematically optimizing primary metabolism,[33] thereby surpassing oleochemical production in *E. coli* for the first time. This high-performance FA platform can produce not only FFAs, but also FALs and alkanes. However, for both *E. coli* and *S. cerevisiae*, the titers were still below commercial levels, and the theoretical yield has never been attained. Low yield and productivity are the key problems hampering industrial application of FA-derived biofuels by microbial fermentation. Therefore, fundamentally new approaches are warranted in order to systematically identify and quantitatively understand all the factors that control carbon flux to FAs. One challenge is to balance the expression of upstream and downstream genes at optimal levels to ensure a balanced supply of precursors and to prevent accumulation of toxic intermediates. Recently, Cheong *et al.* successfully combined non-decarboxylative Claisen condensation reactions and subsequent β-reduction reactions *in vivo* to convert a substrate to the desired functionalized small molecules in an automated engineering workflow.[70,71] Additionally, with the concept of a microbial consortium, a cocultivation approach could be further developed to improve the yield and productivity of target strains with other microorganisms that produce and secrete precursors from inexpensive carbon sources. The central aim of metabolic engineering is to efficiently optimize, manipulate, and utilize useful biological features by defining the metabolic compositions, pathways, and networks and the interactions between organisms and the environment.

Currently, bioethanol and biodiesel are dominating the current global market, even though they have some disadvantages in the form of undesirable physical properties. Among various biofuels, ideal biofuels are bio-hydrocarbons, especially the medium- to long-chain fatty alkanes or alkenes that mimic the chemical composition and physical characteristics of petroleum-based fuels. However, these long-chain hydrocarbons require further chemical processing before being used as fuel. The issue of biodiesel and renewable diesel is complex, because there are advantages and disadvantages to either in terms of fuel properties, environmental issues, and energy balance.[72] The development of pathways to expand the source of feedstock and to tune the properties of biofuel products, additional synthetic biology approaches to optimize microbial hosts, and creation of additional metabolic engineering techniques to improve pathway flux would allow commercialization of advanced biofuels in the immediate future.

References

1. Peralta-Yahya, P.P., Zhang, F., del Cardayre, S.B., & Keasling, J.D. Microbial engineering for the production of advanced biofuels. *Nature* **488**, 320–328 (2012).
2. Campbell, J.W., & Cronan, J.E.J. Bacterial fatty acid biosynthesis: Targets for antibacterial drug discovery. *Annu. Rev. Microbiol.* **55**, 305–332 (2001).
3. Wang, J. *et al*. Discovery of platencin, a dual FabF and FabH inhibitor with *in vivo* antibiotic properties. *Proc. Natl. Acad. Sci. USA* **104**, 7612–7616 (2007).
4. Schweizer, E., & Hofmann, J. Microbial type I fatty acid synthases (FAS): Major players in a network of cellular FAS systems. *Microbiol. Mol. Biol. Rev.* **68**, 501–517 (2004).
5. Rock, C.O., & Cronan, J.E. *Escherichia coli* as a model for the regulation of dissociable (type II) fatty acid biosynthesis. *Biochim. Biophys. Acta.* **1302**, 1–16 (1996).
6. Lomakin, I.B., Xiong, Y., & Steitz, T.A. The crystal structure of yeast fatty acid synthase, a cellular machine with eight active sites working together. *Cell* **129**, 319–332 (2007).
7. White, S.W., Zheng, J., Zhang, Y.-M., & Rock, C.O. The structural biology of type II fatty acid biosynthesis. *Annu. Rev. Biochem.* **74**, 791–831 (2005).
8. Liu, T., Vora, H., & Khosla, C. Quantitative analysis and engineering of fatty acid biosynthesis in *E. coli. Metab. Eng.* **12**, 378–386 (2010).
9. Cronan, J.E., & Thomas, J. Bacterial fatty acid synthesis and its relationships with polyketide synthetic pathways. *Methods Enzymol.* **459**, 395–433 (2009).
10. Zhang, Y.-M., & Rock, C.O. Transcriptional regulation in bacterial membrane lipid synthesis. *J. Lipid Res.* **50**, Suppl, S115–S119 (2009).

11. Fujita, Y., Matsuoka, H., & Hirooka, K. Regulation of fatty acid metabolism in bacteria. *Mol. Microbiol.* **66**, 829–839 (2007).
12. Davis, M.S., & Cronan, J.E.J. Inhibition of *Escherichia coli* acetyl coenzyme A carboxylase by acyl-acyl carrier protein. *J. Bacteriol.* **183**, 1499–1503 (2001).
13. Leibundgut, M., Maier, T., Jenni, S., & Ban, N. The multienzyme architecture of eukaryotic fatty acid synthases. *Curr. Opin. Struct. Biol.* **18**, 714–725 (2008).
14. van Roermund, C.W.T. *et al.* Peroxisomal fatty acid uptake mechanism in *Saccharomyces cerevisiae*. *J. Biol. Chem.* **287**, 20144–20153 (2012).
15. Tehlivets, O., Scheuringer, K., & Kohlwein, S.D. Fatty acid synthesis and elongation in yeast. *Biochim. Biophys. Acta.* **1771**, 255–270 (2007).
16. Lian, J., & Zhao, H. Recent advances in biosynthesis of fatty acids derived products in *Saccharomyces cerevisiae* via enhanced supply of precursor metabolites. *J. Ind. Microbiol. Biotechnol.* **42**, 437–451 (2015).
17. Gronenberg, L.S., Marcheschi, R.J., & Liao, J.C. Next generation biofuel engineering in prokaryotes. *Curr. Opin. Chem. Biol.* **17**, 462–471 (2013).
18. Li, X. *et al.* Overproduction of fatty acids in engineered *Saccharomyces cerevisiae*. *Biotechnol. Bioeng.* **111**, 1841–1852 (2014).
19. Yu, X., Liu, T., Zhu, F., & Khoslaa, C. In vitro reconstitution and steady-state analysis of the fatty acid synthase from *Escherichia coli*. *Proc. Natl. Acad. Sci. USA* **108**, 18643–18648 (2011).
20. Lu, X., Vora, H., & Khosla, C. Overproduction of free fatty acids in *E. coli*: Implications for biodiesel production. *Metab. Eng.* **10**, 333–339 (2008).
21. Zhang, F. *et al.* Enhancing fatty acid production by the expression of the regulatory transcription factor FadR. *Metab. Eng.* **14**, 653–660 (2012).
22. Xu, P. *et al.* Modular optimization of multi-gene pathways for fatty acids production in *E. coli*. *Nat. Commun.* **4**, 1409 (2013).
23. Dellomonaco, C., Clomburg, J.M., Miller, E.N., & Gonzalez, R. Engineered reversal of the beta-oxidation cycle for the synthesis of fuels and chemicals. *Nature* **476**, 355–359 (2011).
24. Juminaga, D. *et al.* Modular engineering of L-tyrosine production in *Escherichia coli*. *Appl. Environ. Microbiol.* **78**, 89–98 (2012).
25. Anthonya, J.R. *et al.* Optimization of the mevalonate-based isoprenoid biosynthetic pathway in *Escherichia coli* for production of the anti-malarial drug precursor amorpha-4, 11-diene. *Metab. Eng.* **11**, 13–19 (2009).
26. Ajikumar, P.K. *et al.* Isoprenoid pathway optimization for Taxol precursor overproduction in *Escherichia coli*. *Science* **330**, 70–74 (2010).

27. Xu, P., Li, L., Zhanga, F., Stephanopoulos, G., & Koffas, M. Improving fatty acids production by engineering dynamic pathway regulation and metabolic control. *Proc. Natl. Acad. Sci. USA* **111**, 11299–11304 (2014).

28. Borodina, I., & Nielsen, J. Advances in metabolic engineering of yeast *Saccharomyces cerevisiae* for production of chemicals. *Biotechnol. J.* **9**, 609–620 (2014).

29. Peralta-Yahya, P.P. *et al.* Identification and microbial production of a terpene-based advanced biofuel. *Nat. Commun.* **2**, 483 (2011).

30. Matsuda, F. *et al.* Increased isobutanol production in *Saccharomyces cerevisiae* by eliminating competing pathways and resolving cofactor imbalance. *Microb. Cell Fact.* **12**, 119–129 (2013).

31. Chen, X., Nielsen, K.F., Borodina, I., Kielland-Brandt, M.C., & Karhumaa, K. Increased isobutanol production in *Saccharomyces cerevisiae* by overexpression of genes in valine metabolism. *Biotechnol. Biofuels* **4**, 21–32 (2011).

32. Yu, A.Q., Juwono, N.K.P., Foo, J.L., Leong, S.S.J., & Chang, M.W. Metabolic engineering of *Saccharomyces cerevisiae* for the overproduction of short branched-chain fatty acids. *Metab. Eng.* **34**, 36–43 (2016).

33. Zhou, Y.J. *et al.* Production of fatty acid-derived oleochemicals and biofuels by synthetic yeast cell factories. *Nat. Commun.* **7**, 11709 (2016).

34. Stephanopoulos, G. Challenges in engineering microbes for biofuels production. *Science* **315**, 801–804 (2007).

35. Fortman, J.L. *et al.* Biofuel alternatives to ethanol: Pumping the microbial well. *Trends Biotechnol.* **26**, 375–381 (2008).

36. Teerawanichpan, P., Robertson, A.J., & Qiu, X. A fatty acyl-CoA reductase highly expressed in the head of honey bee (Apis mellifera) involves biosynthesis of a wide range of aliphatic fatty alcohols. *Insect Biochem. Mol. Biol.* **40**, 641–649 (2010).

37. Akhtar, M.K., Turner, N.J., & Jones, P.R. Carboxylic acid reductase is a versatile enzyme for the conversion of fatty acids into fuels and chemical commodities. *Proc. Natl. Acad. Sci. USA* **110**, 87–92 (2013).

38. Schirmer, A., Rude, M.A., Li, X., Popova, E., & del Cardayre, S.B. Microbial biosynthesis of alkanes. *Science* **329**, 559–562 (2010).

39. Liu, R. *et al.* Metabolic engineering of fatty acyl-ACP reductase-dependent pathway to improve fatty alcohol production in *Escherichia coli. Metab. Eng.* **22**, 10–21 (2014).

40. Teerawanichpan, P., & Qiu, X. Fatty acyl-CoA reductase and wax synthase from *Euglena gracilis* in the biosynthesis of medium-chain wax esters. *Lipids* **45**, 263–273 (2010).

41. Metz, J.G., Pollard, M.R., Anderson, L., Hayes, T.R., & Lassner, M.W. Purification of a jojoba embryo fatty acyl-coenzyme A reductase and expression of its cDNA in high erucic acid rapeseed. *Plant Physiol.* **122**, 635–644 (2000).

42. Reiser, S., & Somerville, C. Isolation of mutants of *Acinetobacter calcoaceticus* deficient in wax ester synthesis and complementation of one mutation with a gene encoding a fatty acyl coenzyme A reductase. *J. Bacteriol.* **179**, 2969–2975 (1997).

43. Tan, X. *et al.* Photosynthesis driven conversion of carbon dioxide to fatty alcohols and hydrocarbons in cyanobacteria. *Metab. Eng.* **13**, 169–176 (2011).

44. Post-Beittenmiller, D. Biochemistry and molecular biology of wax production in plants. *Annu. Rev. Plant Physiol. Plant Mol. Biol.* **47**, 405–430 (1996).

45. Beller, H.R., Goh, E.-B., & Keasling, J.D. Genes involved in long-chain alkene biosynthesis in *Micrococcus luteus*. *Appl. Environ. Microbiol.* **76**, 1212–1223 (2010).

46. Choi, Y.J., & Lee, S.Y. Microbial production of short-chain alkanes. *Nature* **502**, 571–574 (2013).

47. Howard, T.P. *et al.* Synthesis of customized petroleum-replica fuel molecules by targeted modification of free fatty acid pools in *Escherichia coli*. *Proc. Natl. Acad. Sci. USA* **110**, 7636–7641 (2013).

48. Rude, M.A. *et al.* Terminal olefin (1-alkene) biosynthesis by a novel p450 fatty acid decarboxylase from Jeotgalicoccus species. *Appl. Environ. Microbiol.* **77**, 1718–1727 (2011).

49. Sukovich, D.J., Seffernick, J.L., Richman, J.E., Gralnick, J.A., & Wackett, L.P. Widespread head-to-head hydrocarbon biosynthesis in bacteria and role of OleA. *Appl. Environ. Microbiol.* **76**, 3850–3862 (2010).

50. Cintolesi, A., Clomburg, J.M., & Gonzalez, R. In silico assessment of the metabolic capabilities of an engineered functional reversal of the β-oxidation cycle for the synthesis of longer-chain (C≥4) products. *Metab. Eng.* **23**, 100–115 (2014).

51. Liu, Y. *et al.* Hydrogen peroxide-independent production of α-alkenes by OleTJE P450 fatty acid decarboxylase. *Biotechnol. Biofuels* **7**, 28–40 (2014).

52. Liu, Q. *et al.* Engineering an iterative polyketide pathway in *Escherichia coli* results in single-form alkene and alkane overproduction. *Metab. Eng.* **28**, 82–90 (2015).

53. Zhang, F., Rodriguez, S., & Keasling, J.D. Metabolic engineering of microbial pathways for advanced biofuels production. *Curr. Opin. Chem. Biol.* **22**, 775–783 (2011).

54. Kalscheuer, R., Uthoff, S., Luftmann, H., & Steinbüchel, A. *In vitro* and *in vivo* biosynthesis of wax diesters by an unspecific bifunctional wax ester synthase/acyl-CoA: Diacylglycerol acyltransferase from *Acinetobacter calcoaceticus* ADP1. *Eur. J. Lipid Sci. Technol.* **105**, 578–584 (2003).
55. Stöveken, T., Kalscheuer, R., Malkus, U., Reichelt, R., & Steinbüchel, A. The wax ester synthase/acyl coenzyme A:diacylglycerol acyltransferase from *Acinetobacter* sp. strain ADP1: Characterization of a novel type of acyltransferase. *J. Bacteriol.* **187**, 1369–1376 (2005).
56. Kalscheuer, R., Stölting, T., & Steinbüchel, A. Microdiesel: *Escherichia coli* engineered for fuel production. *Microbiol.* **152**, 2529–2536 (2006).
57. Steen, E.J. *et al.* Microbial production of fatty-acid-derived fuels and chemicals from plant biomass. *Nature* **463**, 559–562 (2010).
58. Duan, Y., Zhu, Z., Cai, K., Tan, X., & Lu, X. De novo biosynthesis of biodiesel by *Escherichia coli* in optimized fed-batch cultivation. *PLoS One* **6**, e20265 (2011).
59. Zhang, F., Carothers, J.M., & Keasling, J.D. Design of a dynamic sensor-regulator system for production of chemicals and fuels derived from fatty acids. *Nat. Biotechnol.* **30**, 354–359 (2012).
60. Layton, D.S., & Trinh, C.T. Engineering modular ester fermentative pathways in *Escherichia coli*. *Metab. Eng.* **26**, 77–88 (2014).
61. Tai, Y.S., Xiong, M., & Zhang, K. Engineered biosynthesis of medium-chain esters in *Escherichia coli*. *Metab. Eng.* **27**, 20–28 (2015).
62. Thompson, R.A., & Trinh, C.T. Enhancing fatty acid ethyl ester production in *Saccharomyces cerevisiae* through metabolic engineering and medium optimization. *Biotechnol. Bioeng.* **111**, 2200–2208 (2014).
63. Shi, S., Valle-Rodríguez, J.O., Khoomrung, S., Siewers, V., & Nielsen, J. Functional expression and characterization of five wax ester synthases in *Saccharomyces cerevisiae* and their utility for biodiesel production. *Biotechnol. Biofuels* **5**, 7 (2012).
64. Shi, S., Valle-Rodríguez, J.O., Siewers, V., & Nielsen, J. Engineering of chromosomal wax ester synthase integrated *Saccharomyces cerevisiae* mutants for improved biosynthesis of fatty acid ethyl esters. *Biotechnol. Bioeng.* **111**, 1740–1747 (2014).
65. Rodriguez, G.M., Tashiro, Y., & Atsumi, S. Expanding ester biosynthesis in *Escherichia coli*. *Nat. Chem. Biol.* **10**, 259–265 (2014).
66. Guo, D., Zhu, J., Deng, Z., & Liu, T. Metabolic engineering of *Escherichia coli* for production of fatty acid short-chain esters through combination of the fatty acid and 2-keto acid pathways. *Metab. Eng.* **22**, 69–75 (2014).

67. Tao, H., Guo, D., Zhang, Y., Deng, Z., & Liu, T. Metabolic engineering of microbes for branched-chain biodiesel production with low-temperature property. *Biotechnol. Biofuels* **8**, 92 (2015).
68. Agnew, D.E., Stevermer, A.K., Youngquist, J.T., & Pfleger, B.F. Engineering *Escherichia coli* for production of C_{12}–C_{14} polyhydroxyalkanoate from glucose. *Metab. Eng.* **14**, 705–713 (2012).
69. Tappel, R.C., Wang, Q., & Nomura, C.T. Precise control of repeating unit composition in biodegradable poly(3-hydroxyalkanoate) polymers synthesized by *Escherichia coli*. *J. Biosci. Bioeng.* **113**, 480–486 (2012).
70. Cheong, S., Clomburg, J.M., & Gonzalez, R. Energy- and carbon-efficient synthesis of functionalized small molecules in bacteria using non-decarboxylative Claisen condensation reactions. *Nat. Biotechnol.* **34**, 556–561 (2016).
71. Ng, C.Y., Chowdhury, A., & Maranas, C.D. A microbial factory for diverse chemicals. *Nat. Biotechnol.* **34**, 513–515 (2016).
72. Knothe, G. Biodiesel and renewable diesel: A comparison. *Prog. Energ. Combust.* **36**, 364–373 (2010).

Chapter 5

Photosynthesis and Its Metabolic Engineering Applications

Jason T. Ku* and Ethan I. Lan[†,‡]

*Institute of Molecular Medicine and Bioengineering
National Chiao Tung University, Hsinchu, Taiwan
†Department of Biological Science and Technology
National Chiao Tung University, Hsinchu, Taiwan
‡ethanilan@nctu.edu.tw

5.1. Introduction of Photosynthesis

Photosynthesis is used by plants, algae, and many microbes to convert light energy into chemical energy, usually resulting in the synthesis of carbohydrates. In a macroscopic view, this process plays an important role on recycling carbon from atmospheric CO_2 and transforming solar energy into chemical energy stored in organic carbon. Photosynthesis consists of light-dependent reaction (light reaction) and light-independent reaction (dark reaction), which produces oxygen and carbohydrates, respectively. Light reaction, as the name implies, begins with photosystem II (PSII) absorbing energy from photons and subsequently using it to split water into O_2, H^+, and electrons. The excited electrons pass through a series of electron-transferring redox reactions, eventually reaching plastocyanin in the thylakoid lumen. Plastocyanin then passes the electrons to photosystem I (PSI), which undergoes another photon-dependent excitation. The excited electrons eventually end up at ferredoxin in the stroma of chloroplast or cytosol in cyanobacteria. These reduced ferredoxins can then transfer the electrons to $NADP^+$, forming the primary reducing equivalent NADPH for photosynthetic organisms (Figure 5.1). This process of the light reaction is frequently referred to as the Z-scheme. During the process of electron transfer in the Z-scheme, protons are pumped across the thylakoid membrane, effectively forming an electrochemical potential. This proton gradient is then used by ATP synthase to generate ATP for cellular functions.

Dark reaction, on the other hand, uses the energy generated by the light reaction to fix CO_2 into central metabolites. For the majority of photosynthetic organisms, the Calvin–Benson–Bassham (CBB) cycle is used for carbon fixation (Figure 5.1). With the exception of ribulose-1,5-bisphosphate

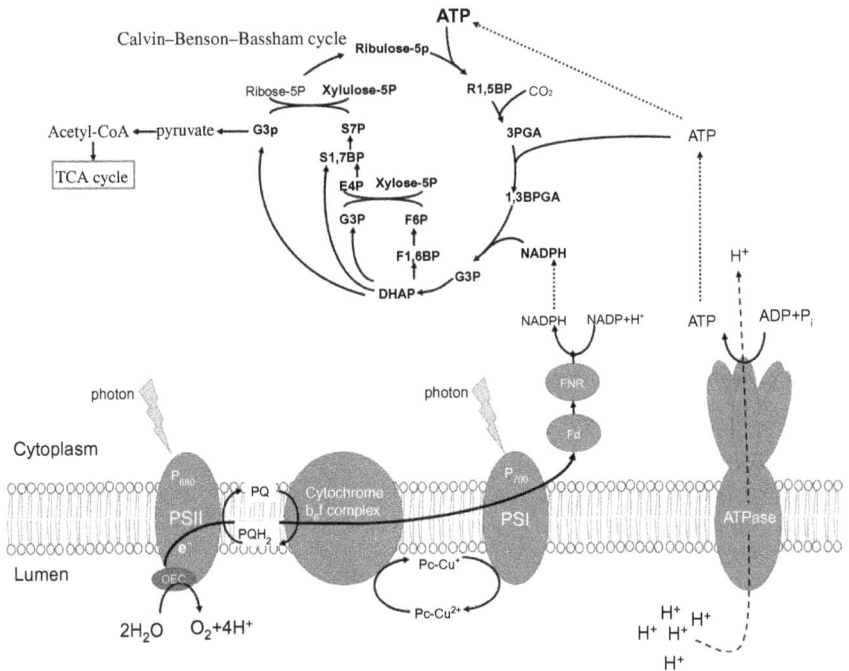

Figure 5.1. Schematics of photosynthetic machinery and the CBB cycle. In light reaction, the photons are absorbed by PSII and PSI to excite the electrons, which pass through Z-scheme and are finally accepted by $NADP^+$. The forming NADPH and ATP in light reaction are then used in CBB cycle to fix CO_2 into central metabolites.

carboxylase/oxygenase (RuBisCo), catalyzing the fixation of CO_2 into 3-phosphoglycerate and sedoheptulose bisphosphatase, catalyzing the dephosphorylation of sedoheptulose 1,7-bisphosphate, and phosphoribulokinase, and catalyzing the generation of ribulose-1,5-bisphosphate for RuBisCo reaction, all other enzymes of the CBB cycle are present in most heterotrophs in the pentose phosphate pathway. The net reaction of the CBB cycle is the synthesis of a glyceraldehyde-3-phosphate from 3 CO_2, 6 NADPH, and approximately 9 ATP. In addition to the CBB cycle, other CO_2 fixation pathways have been identified in nature. These pathways include the reductive tricarboxylic acid (TCA) cycle, 3-hydroxypropionate bicycle, and the reductive acetyl-CoA pathway. These pathways differ from CBB cycle in the amount of ATP required and reducing equivalent utilized.

Increasingly, concerns over the depletion of fossil resources and the increase of atmospheric CO_2 have driven rapid research advancement in biochemical and biofuel production. While conventional sugar-based fermentations have been successful, recent attention has been shifted toward the utilization of non-food-based resources including various wastes. Among these non-conventional resources, CO_2 is one of the most studied carbon source. Utilizing photosynthetic organisms for biochemical and biofuel production is an attractive direction for sustainability because it bypasses the need of repeated growth and deconstruction of plant biomass. In particular, the metabolic engineering of cyanobacteria has gained significant attention in recent days. Cyanobacteria have well-developed genetic tools, which enable them to be rapidly engineered and modified for desirable phenotypes including the production of biochemicals and biofuels. Among all cyanobacteria, *Synechocystis* PCC6803, *Synechococcus* sp. PCC7002, *Anabaena* sp. PCC7120, and *Synechococcus elongatus* PCC7942 (hereafter denoted PCC6903, PCC7002, PCC7120, and PCC7942, respectively) are the most commonly used model strains of because their well-studied genetic and genomic information and their ease of manipulation. PCC6803 and PCC7942 were the first fully sequenced cyanobacteria strains and the first demonstrated transformable by foreign DNA. Both PCC6803 and PCC7942 are freshwater cyanobacteria.[1,2] On the other hand, PCC7002 is a marine species, which has advantage of higher salinity and light-intensity tolerance.[3,4] PCC7120 is a diazotrophic capable of nitrogen fixation, which reduces the need for exogenous nitrogen source. With the exception of PCC7120, all other model strains are naturally competent, which facilitates the ease of genetic and metabolic engineering. Recently a rapid-growing cyanobacterium *Synechococcus elongatus* UTEX 2973 was isolated.[5] This strain of *S. elongatus* has genomic structure very similar to that of PCC7942. However, the growth rate of this strain is reported to be twice that of PCC7942, making it a more favorable host for cyanobacterial chemical production. In this chapter, the advances on cyanobacterial metabolic engineering are covered. In particular, this chapter focuses on the development of cyanobacterial strains for the production of various chemicals including hydroxy acids, alcohols, olefins, sugars, and more. Particular emphasis will be placed on the metabolic engineering strategies used in each case.

5.2. Recent Metabolic Engineering of Cyanobacteria for Chemical Synthesis

Metabolic engineering of cyanobacteria for the photosynthetic production of biochemicals began in the late 20th century. With increasing knowledge of genomic information and genetic tools, the number of cyanobacterial biochemical productions rapidly increased in the past decade. In most recombinant cyanobacteria strain development, homologous recombination was used instead of plasmids for gene overexpression. To drive heterologous genes, IPTG-inducible promoters, including PLac, PTrc, and PTac, and constitutive promoters such as psbA2 and rbcL are frequently used. The IPTG-inducible promoters are the same as those commonly used in *Escherichia coli* engineering and offers control of gene overexpression at various stages of growth. On the other hand, the constitutive promoters PpsbA2 and PrbcL are natively used to drive photosynthetic machinery in cyanobacteria. These promoters are selected for their high level of expression. Table 5.1 summarizes the current state of chemicals produced by engineered cyanobacteria, with emphasis on genes overexpressed, promoters used, and titers, and will be used as a reference for the discussions covered in this chapter. As can be seen in Table 5.1, cyanobacterial biochemical production titers range from microgram to gram per liter. Currently, the productions that reach the grams per liter level are considered relatively higher flux photosynthetic biochemical productions.

Table 5.1. Chemicals and genetic information of engineered cyanobacteria.*

Chemical target	Strain	Relevant genotypes	Titer (mg/L)	Culture time (day)	Relevant comments	References
Ethanol	PCC7942	P_{rbcLS}::*pdc,adh*	80	21	First demonstration of photosynthetic ethanol production. Expression of *pdc* and *adh* on a plasmid	[6]
Ethanol	PCC6803	P_{psbA2}::*pdc,adh*	550	6	Strong light-driven promoter and production via photobioreactor	[7]
Ethanol	PCC6803	P_{rbc}::*pdc,adh(slr1192)*, Δphb	5,500	26	Increased gene dosage by expressing two copies of *pdc* and *adh*. Culture condition optimization	[8]

(Continued)

Table 5.1. (*Continued*)

Chemical target	Strain	Relevant genotypes	Titer (mg/L)	Culture time (day)	Relevant comments	References
Ethanol	PCC7942	P$_{trc}$::*pduP,yqhD*	182	10	Expression of oxygen-tolerant CoA-acylating aldehyde dehydrogenase and alcohol dehydrogenase	[9]
Isopropanol	PCC7942	P$_{LlacO1}$::*thl,atoAD,adc* P$_{LlacO1}$::*adh*	26.5	9	Isopropanol was only observed upon dark anoxic incubation	[10]
Isopropanol	PCC7942	P$_{LlacO1}$::*thl,atoAD,adc* P$_{LlacO1}$::*adh*	146	10	Dark anoxic incubation with optimized production condition	[11]
Isopropanol	PCC7942	P$_{LlacO1}$::*thl, atoAD, adc, sadh* P$_{LlacO1}$::*pta*	33.1	14	Production under photosynthetic conditions	[12]
1-Butanol	PCC7942	P$_{trc}$::*ter* P$_{LlacO1}$::*atoB,adhE2,crt,hbd*	14.5	7	1-Butanol was only observed upon dark anoxic incubation	[13]

Product	Strain	Genes				Ref
1-Butanol	PCC7942	$P_{trc}::ter$ $P_{LlacO1}::nphT7,bldh,yqhD,crt,hbd$	29.9	17	Use of ATP driving force and NADPH-dependent enzymes enabled first demonstration of butanol production under standard photosynthetic conditions	[14]
1-Butanol	PCC7942	$P_{trc}::ter$ $P_{LlacO1}::nphT7,pduP,yqhD,phaJ,phaB$	404	12	Using oxygen tolerance enhanced productivity 10-fold	[9]
Isobutanol	PCC7942	$P_{trc}::kivd,yqhD$ $P_{LlacO1}::alsS,ilvC,ilvD$	450	6	Decarboxylations serve as a driving force	[15]
Isobutanol	PCC6803	$P_{tac}::kivD,adhA$	240	21	*In situ* removal products by oleyl alcohol trap	[16]
Isobutyraldehyde	PCC7942	$P_{trc}::kivd/P_{LlacO1}::alsS,ilvC,ilvD$ $P_{tac}::rbcL,rbcS$	1,100	8	Decarboxylations serve as a driving force. Gas stripping provided *in situ* product removal	[15]

(Continued)

Table 5.1.   (*Continued*)

Chemical target	Strain	Relevant genotypes	Titer (mg/L)	Culture time (day)	Relevant comments	References
2-Methyl-1-butanol	PCC7942	P_{trc}::*kivd,yqhD* P_{trc}::*cimAΔ2,leuBCD*	177.5	12	Native isoleucine pathway shown highly active	[17]
Glycerol	PCC6803	P_{trc}::*gpp2*	1,315	17	Mild salt stress applied	[18]
Glycerol	PCC7942	P_{trc}::*gpp1*	1,170	20	Standard photosynthetic CO_2 bubbling conditions	[19]
Glycerol	PCC7942	P_{LlacO1}::*gpd1,hor2*	1,160	14	Standard photosynthetic CO_2 bubbling conditions	[20]
2,3-Butanediol	PCC6803	P_{trc}::*als,aldc,ar*	430	29	Codon-optimized genes, product was mixed in chirality	[21]
2,3-Butanediol	PCC7942	P_{LlacO1}::*alsS,alsD,adh*	2,380	21	Showed that 2,3-butanediol has low cellular toxicity	[22]
1,2-Propanediol	PCC7942	P_{trc}::*sadh,yqhD,mgsA*	150	10	Expression of NADPH-utilizing enzyme	[23]

1,3-Propanediol	PCC7942	P_{LlacO1}::dhaB123,gdrAB,yqhD P_{LlacO1}::gpd1,hor2	288	14	Demonstrated that no external vitamin B12 was needed	[20]
1,3-Propanediol	PCC7942	P_{trc}::dhaB123,gdrAB,yqhD P_{LlacO1}::gpd1,hor2	1220	20	2-fold phosphate level and high IPTG level	[24]
Erythritol	PCC6803	P_{psbA2}::pktS7942 P_{cpcBA}::e4p(tm1254)/ P_{trc}::gld1	256	28	First erythritol production in non-native organism	[25]
Mannitol	PCC7002	P_{psbA}::mtlD,mlp,ΔglgA1,ΔglgA2	1,100	12	Expression of codon-optimized genes	[26]
Fatty acid	PCC6803	P_{cpc}::accBC/P_{thc}::accDA,fatB2 P_{trc}::tesA,fatB1,fatB2,Δpta P_{psbA2}::fatB1,fatB2,Δaas,Δphp2	197	2	Codon optimized, six-generation strain construction including reverse and competing pathway knocked out	[27]
Fatty acid	PCC6803	Δaas	17.5	10	Knocked out fatty-acyl-ACP synthetase	[28]
Fatty acid	PCC7942	P_{trc}::tesA, Δaas	45	20	Knocked out gene for fatty acid degradation	[29]

(Continued)

Table 5.1. (*Continued*)

Chemical target	Strain	Relevant genotypes	Titer (mg/L)	Culture time (day)	Relevant comments	References
Fatty acid	PCC7942	P_{trc}::fat1,rbcLS, Δaas	30	20	ACCase increased fatty acid production on a DCW basis	[30]
Fatty acid	PCC7002	P_{trc}::tesA/P_{psbA1}::rbcLS,ΔfadD	131	20	Overexpressed RuBisCO subunits. Shown that fatty acid tolerance depends on temperature	[31]
Fatty acid ethyl ester	PCC7942	P_{trc}::aftA,xpkA,pta/P_{trc}::pdc,adh			10 mg/L/OD, first demonstration	[32]
Fatty alcohol	PCC6803	P_{rbc}::far	0.2	18	Uses native fatty-acyl-ACP synthesis	[33]
Fatty alcohol	PCC6803	P_{rbc}::far/P_{psbA2}::aas	0.17	10	Overexpressed fatty-acyl-ACP synthesis	[28]
Fatty alcohol	PCC6803	P_{petE}::maqu_2220,Δsll0208, Δsll0209	2.1	16	Fatty-acyl-CoA reductase used and knocked out alkane biosynthetic pathway	[34]

D-Lactate	PCC7942	P_{lac}::*ldhA,lldP,udhA*	55	4	Expression of transporter shown to help lactate efflux	[35]
D-Lactate	PCC7942	P_{trc}::*ldhD,lldP*	829	10	Engineered LdhD to increase NADPH utilization	[18]
D-Lactate	PCC6803	P_{trc}::*gldA101,sth*	1,140	24	Codon-optimized GldA. Acetate addition increased titer	[36]
L-Lactate	PCC6803	P_{psbA2}::*ldh,lldP*	15.3	18	Tested different Ldh	[37]
L-Lactate	PCC6803	P_{trc2}::*pk*/P_{trc2}::*ldh*	837	14	Ppc knocked down	[38]
L-Lactate	PCC6803	P_{trc}::*ldh*	1,800	30	Codon optimization and long-term production	[39]
L-Lactate	PCC6803	P_{trc}::*ldh*/P_{trc}::*sth*	288	14	Expression of transhydrogenase led to growth retardation. Expression of *ldh* rescued the phenotype	[40]

(*Continued*)

Table 5.1. *(Continued)*

Chemical target	Strain	Relevant genotypes	Titer (mg/L)	Culture time (day)	Relevant comments	References
3-Hydroxypropionate	PCC7942	P_{trc}::msr,mcr P_{LlacO1}::adc,Skpyd4,ppc,aspC	665	12	Combination of malonyl-CoA dependent and β-alanine pathway	[41]
3-Hydroxypropionate	PCC6803	P_{cpc560}::mcr,pntAB P_{petE5}::accBCAD,birA	837	8	Increased malonyl-CoA supply and NADPH availability	[42]
3-Hydroxypropionate	PCC7942	P_{trc}::gpp1,puuC	31.7	10	Phosphate and nitrogen-limitation	[19]
3-Hydroxybutyrate	PCC6803	P_{tac}::tesB/P_{tac}::phaAB,ΔphaEC	533.4	21	Continuous production from atmosphere CO_2; Productivity increased after cell growth reached stationary phase	[43]
Succinic acid	PCC6803	P_{psbA2}::sigE,Δack	113	4	Expressing sigma factor	[44]

Succinic acid	PCC6803	$P_{trc}::ppc, \Delta slr0168$	192	3	Expressing limiting step enzyme PPC	[45]
Succinic acid	PCC7942	$P_{trc}:: ppc,gltA, \Delta glgc$	0.44	2	Apply CRISPE-Cas9 mechanism, nitrate-free	[46]
Succinic acid	PCC7942	$P_{trc}::gabD,kgd,gltA,ppc$	430	8	Succinic acid production under normal photosynthesis condition	[47]
p-Coumaric acid	PCC6803	$P_{psbA2}::sam8, \Delta slr1573$	82.6	4	*p*-Coumarate only observed after deletion of laccase gene	[48]
Caffeic acid	PCC6803	$P_{psbA2}::sref8$	7.2	3	First demonstration of photosynthetic production of caffeic acid	[49]
Itaconic acid	PCC6803	$P_{tac}::cad$	14.5	16	First demonstration of photosynthetic itaconic acid production	[50]

(Continued)

Table 5.1.   (*Continued*)

Chemical target	Strain	Relevant genotypes	Titer (mg/L)	Culture time (day)	Relevant comments	References
Ethylene	PCC7942	P_{psbA1}::*efe*			10.82 mL/L/D/OD	[51]
Ethylene	PCC6803	P_{psbA}::*efe*			171 mg/L/day, multiple *efe* expressed	[52]
Ethylene	PCC6803	P_{psbA}::*efe,Δslr0168*			17.2 mL/L/D/OD	[53]
Ethylene	PCC7942	P_{trc}::ACS-Ctdoc,ACO-Acdoc,Cip			81.6 mL/L/D/OD, different induction level	[54]
Ethylene	PCC6803	P_{trc}::*efe*			240 mL/L/D, various promoter tested	[55]
Isoprene	PCC6803	P_{psbA2}::*IspS*			50 µg/g CDW, codon optimized	[56]
Isoprene	PCC6803	P_{psbA2}::*IspS*	0.35	8	Gaseous/aqueous two-phase photobioreactors	[57]
Isoprene	PCC6803	P_{psbA2}::*hmgS,hmgR,atoB* P_{psbA2}:: *fni,mk,pmd,pmk* P_{psbA2}::*Isps*	0.3	8	First to use MVA pathway	[58]

Product	Strain					Reference
Isoprene	PCC7942	P_{trc}::idi-GGGS-*IspS,dxs* P_{tac}::*IspG*	1.26	21	Engineered MEP pathway to increase flux and identified bottleneck as *IspG*	[59]
Limonene	PCC7120	P_{nir}/P_{psbA1}::*lims,dxs,ipphp,gpps*	0.52	12	High light density improved production	[60]
Limonene	PCC7002	P_{cpcBA}::*lims*	4	4	Product trap by dodecane overlay	[61]
Limonene	PCC6803	P_{trc}::*lims*/P_{trc}::*dxs,crt,ipi*	1	30	Constant productivity and MEP pathway genes overexpressed	[62]
Limonene	PCC7942	P_{trc}::*ls*			885.1 μg/L/OD/d, change RBS improved productivity	[63]
Bisabolene	PCC6803	P_{cpcBA}::*bis*	0.6	4	Dodecane overlay as product trap	[61]
β-Phellandrene	PCC6803	P_{psbA2}::*phls*	0.2	8	Codon-optimized PHLs	[64]

(Continued)

Table 5.1. (*Continued*)

Chemical target	Strain	Relevant genotypes	Titer (mg/L)	Culture time (day)	Relevant comments	References
β-Phellandrene	PCC6803	$P_{cpc}::cpcB\bullet phls.cpcA$			3.2 mg/g DCW, fusion protein	[65]
β-Phellandrene	PCC6803	$P_{psbA2\cdot trc\cdot T7}::phls$	0.9	2	Fused promoter and high light intensity	[66]
β-Caryophyllene	PCC6803	$P_{psbA2}::qhs1$	0.046	7	Suggested β-phellandrene is transcriptionally regulated	[67]
Farnesene	PCC7120	$P_{nir}\text{-}P_{psbA2}::ifas$	0.3	15	Codon optimized	[68]
Squalene	PCC6803	Δshc			0.5 mg/L/OD, knockout squalene hopene cyclase	[69]
Hexose	PCC7942	$P_{trc}::glf,invA,galU$	85	5	Transporter expressed facilitated monosaccharide export	[35]

Product	Strain	Genetic modification				Reference
Sucrose	PCC7942	P_{trc}::cscB, ΔinvA, ΔglgC	2,700	7	Knockout of the major competing pathways	[70]
Sucrose	PCC6803	P_{petE}::cscB,sps,spp,ugp, Δggps, ΔggtCD	140	10	Salt stress	[71]
Sucrose	UTEX 2973	Plac::cscB	3,340	4	Salt stress applied. Seven production cycles of resuspending cells into new media led to 8.7 g/L sucrose produced in cumulative amount	[72]
Glucosylglycerol	PCC6803	ΔggpR, ΔggtCD	981	24	Semicontinuous culturing conditions	[73]
Acetone	PCC6803	P_{rbc}::cftAB-P_{cpc}::adc, ΔphaCE,pta	36	4	Increasing acetyl-CoA pool	[74]

*This table is modified from Lai and Lan.[75]

5.2.1. *Short-Chain Alcohols*

5.2.1.1. *Ethanol*

As one of the most well-known and successful bioproducts, ethanol can also be produced by engineered cyanobacteria through photosynthesis upon the expression of genes for its production pathway. The first report of photosynthetic production of ethanol was published in 1999 where PCC7942 was engineered to express heterologous pyruvate decarboxylase (PDC) and alcohol dehydrogenase (ADH) from *Zymomonas mobilis* (Figure 5.2)[6] on a plasmid. The resulting strain achieved an ethanol titer of around 80 mg/L in 21 days. It was noted that the productivity of ethanol by this recombinant strain was significantly lower than the detected *in vitro* activity of the two enzymes in the pathway, indicating ineffective channeling of carbon flux. Roughly 10 years later, due to increased interest in photosynthetic chemical conversion, recombinant cyanobacteria for ethanol production was again studied. In a 2009 study,[7] PCC6803 was engineered to express the same *pdc* and *adh* from *Z. mobilis*. However, in this study, a strong light-driven promoter P_{psba2} was used drive the synthetic operon. In addition, the genes were homologously recombined into the chromosome of PCC6803, greatly enhancing its stability. The resulting strain produced over 550 mg/L of ethanol using a photobioreactor. Ethanol production was further increased to a titer over 5.5 g/L with daily productivity of 212 mg/L/d.[8] This study also used PCC6803 as the model organism. Different from previous studies, the genes *pdc* and *adh* were duplicated to increase the gene dosage for enzyme overexpression. A native ADH Slr1192 from PCC6803 was used instead of *Z. mobilis* ADH. Furthermore, gene encoding for poly-β-hydroxybutyrate, a major carbon storage for PCC6803, synthesis was knocked out to increase the flux to ethanol.

In addition to the PDC pathway, ethanol can also be synthesized from acetyl-CoA via a two-step reduction. PCC7942 was engineered to express heterologous CoA-acylating aldehyde dehydrogenase PduP from *Salmonella enterica* and an NADPH-dependent ADH YqhD from *E. coli*.[9] Upon introduction of this pathway into PCC7942, the resulting strain produced 182 mg/L of ethanol. While lower than the PDC-mediated pathway, this work showed that ethanol production from acetyl-CoA was

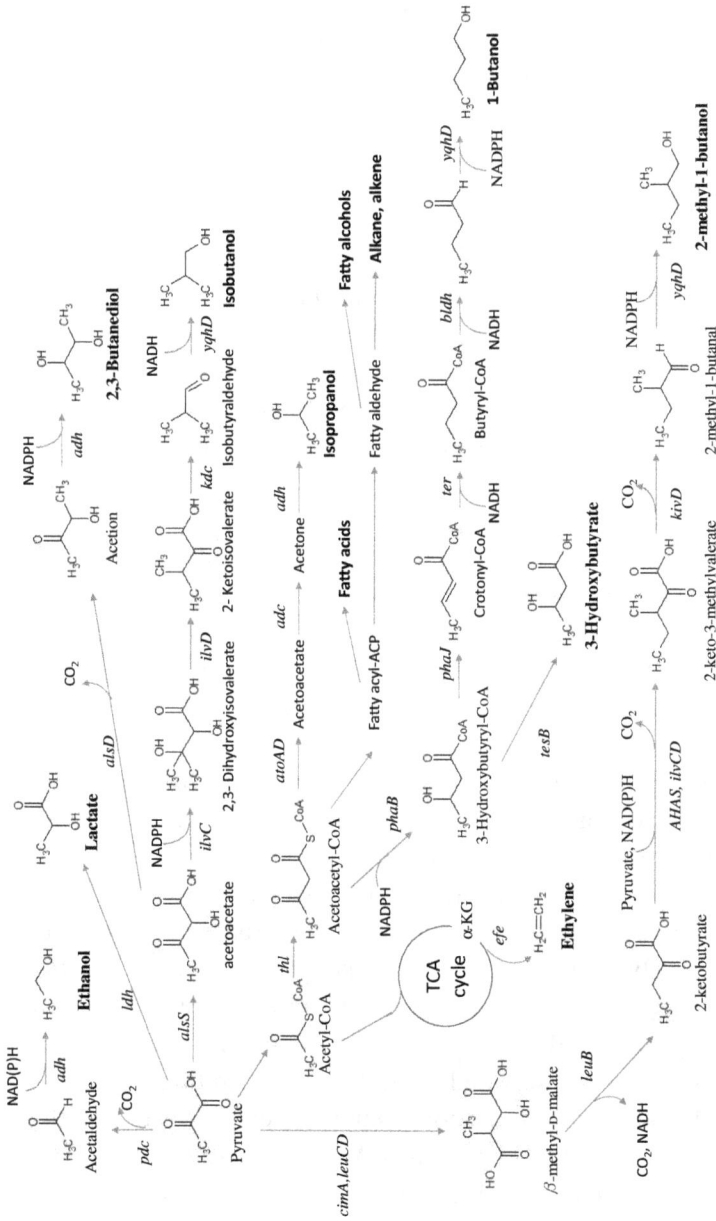

Figure 5.2. Schematics of pyruvate-derived chemical production pathways. Abbreviations: *adh*, aldehyde dehydrogenase; *PDC*, pyruvate decarboxylase; *ldh*, lactate dehydrogenase; *alsS*,α-acetolactate synthase; *alsD*, acetolactate decarboxylase; *ilvC*, acetohydroxy acid isomer oreductase; *ilvD*, dihydroxy-acid dehydratase; *kdc*, 2-ketoacid decarboxylase; *yqhD*, alcohol dehydrogenase; *thl*, thiolase; *atoAD*, CoA transferase; *adc*, acetoacetate decarboxylase; *phaB*, acetoacetyl-CoA reductase; *phaJ*, enoyl-CoA hydratase; *ter*, trans-2-enoyl-CoA reductase; *bldh*, butyraldehyde dehydrogenase; *tesB* thioesterase; *cimA*, citramalate synthase; *leuB*, 3-isopropylmalate dehydrogenase; AHAS, acetohydroxyacid synthase; *kivD*, ketoacid decarboxylase.

possible once an oxygen-tolerant aldehyde dehydrogenase was used. This was a significant advancement as most CoA-acylating aldehyde dehydrogenase naturally used for ethanol production is known to be oxygen sensitive. The importance of this feature will be further discussed in butanol production in Section 5.2.1.3.

5.2.1.2. *Isopropanol*

Isopropanol is a precursor to propylene, which is the monomer for polypropylene, a common polymer widely used in textile, medicine, and plastic flexible hinges such as flip-top bottle. Isopropanol is a fermentation product derived from acetone, naturally produced by some *Clostridia* species. The isopropanol pathway (Figure 5.2) is composed of acetyl-CoA acetyltransferase (encoded by *thl*), acetoacetyl-CoA transferase (encoded by *atoAD*), acetoacetate decarboxylase (encoded by *adc*), and secondary ADH (encoded by *adh*). In this pathway, acetyl-CoA acetyltransferase condenses two acetyl-CoA into acetoacetyl-CoA, which will subsequently be converted to acetoacetate by acetoacetyl-CoA transferase. Acetoacetate is then decarboxylated into acetone, which is reduced to isopropanol. To generate a strain of cyanobacteria capable of producing isopropanol from CO_2, all of these genes were integrated into the genome of PCC7942 via homologous recombination.[10] Contrary to expectations, the resulting strain was not able to produce isopropanol under standard photosynthetic conditions. Small amounts of isopropanol were observed in the medium upon acetate feeding. This observation indicated that under standard conditions, acetate is absent and hinders the function of acetoacetyl-CoA transferase. Additionally, this result may also suggest that carbon flux to acetyl-CoA under photosynthetic conditions may not be sufficient to support isopropanol production. To increase acetyl-CoA flux, this recombinant strain was cultivated under nutrient-limiting conditions and incubated under dark and anoxic environment. Effectively, this strategy forced carbon flux to acetyl-CoA and acetate, resulting in a production of isopropanol with a titer of 26.5 mg/L in 9 days. This production strain was further optimized to produce isopropanol with a titer of 146 mg/L.[11] In a similar study in PCC6803, acetoacetyl-CoA transferase (CtfAB) was expressed to construct a strain capable of producing 36 mg/L of acetone in 4 days under similar conditions. In PCC6803, heterologous thiolase expression was not needed because it has a native thiolase as part of the polyhydroxybutyrate (see Section 5.2.4.3) biosynthesis operon. Since shifting from standard photosynthetic to dark anoxic condition is complicated, they further engineered the strain with exogenous phosphate acetyltransferase (Pta).[12] The strain carrying *pta* gene reported isopropanol production under photosynthetic condition.

5.2.1.3. *1-Butanol*

1-Butanol is a natural fermentation product synthesized by some *Clostridia* species. 1-Butanol has attracted significant attention in the metabolic engineering field in the mid-2000s. Many major research groups have tried to transfer the butanol production pathway from *Clostridia* to more tractable hosts such as *E. coli* and *Saccharomyces cerevisiae*.[76,77] However, the titers of butanol produced by these recombinant hosts were typically only a small fraction of those produced by *Clostridia*. It was identified that a key enzyme in the pathway butyryl-CoA dehydrogenase (Bcd) was inadequately expressed. Two independent studies instead used another class of enzyme called trans-enoyl-CoA reductase (Ter) to replace Bcd and was able to achieve comparable or higher production of butanol in *E. coli* to that of *Clostridia*. The butanol pathway (Figure 5.2) starts from acetyl-CoA. Two acetyl-CoAs condense into an acetoacetyl-CoA, catalyzed by thiolase (Thl or AtoB), 3-hydroxybutyryl-CoA dehydrogenase (Hbd) then reduces acetoacetyl-CoA into 3-hydroxybutyryl-CoA, which is then dehydrated into crotonyl-CoA by Crotonase (Crt). Subsequently, Ter reduces crotonyl-CoA into butyryl-CoA, which is further reduced to 1-butanol by AdhE2. Similar to the engineering of isopropanol-producing strain of cyanobacteria, expression of this 1-butanol-producing pathway into PCC7942 was not successful for direct photosynthetic production of 1-butanol. 1-Butanol was observed upon dark and anoxic incubation.[13] It was suggested that two bottlenecks needed to be solved for photosynthetic 1-butanol production: the lack of strong driving force and the oxygen sensitivity of AdhE2. To increase the thermodynamic favorability of the acetyl-CoA condensation, acetoacetyl-CoA synthase (NphT7) was used instead of thiolase.[14] NphT7 uses malonyl-CoA and acetyl-CoA to synthesize acetoacetyl-CoA in a decarboxylative and irreversible Claisen condensation. Malonyl-CoA is the result of ATP-dependent activation of acetyl-CoA. Thus the net difference between NphT7 and thiolase-mediated synthesis of acetoacetyl-CoA is the net input of ATP hydrolysis energy. Upon the introduction of NphT7 and substitution of NADPH utilizing dehydrogenases, the recombinant strain of *S. elongatus* was able to produce around 30 mg/L of 1-butanol under photosynthetic conditions. NADPH-utilizing enzymes are preferred for cyanobacterial engineering because NADPH is the natural reducing equivalent produced in photosynthesis. Subsequently, the oxygen sensitivity of AdhE2 was addressed by recruiting PduP, a CoA-acylating aldehyde dehydrogenase from the 1,2-propanediol degradation pathway. PduP was assayed and demonstrated oxygen stability.[9] Utilizing PduP in combination with NphT7, the resulting strain produced up to 400 mg/L of 1-butanol in 8 days, representing more than 10-fold improvement compared to the parent strain.

5.2.1.4. *Isobutanol*

Similar to 1-butanol, isobutanol is a better candidate for gasoline substitute than ethanol. Unlike 1-butanol, isobutanol is not known to be produced in significant amounts by any organism. A synthetic pathway (Figure 5.2) based on valine biosynthesis was constructed to produce isobutanol in recombinant organisms.[78] In this pathway, pyruvate is converted to 2-ketoisovalerate, the precursor for valine. Instead of converting 2-ketoisovalerate to valine, it is decarboxylated and reduced to isobutanol. In addition, the first enzyme of the pathway catalyzes a decarboxylation, effectively serving as a driving force to push carbon flux into the isobutanol pathway. This pathway was introduced into PCC7942 and enabled production of 450 mg/L isobutanol in 6 days in shake flasks.[15] The same study also aimed to produce isobutyraldehyde, which is made by omitting the ADH in heterologous expression. As a more volatile compound, isobutyraldehyde is easily removed from production culture by gas stripping. Together with this *in situ* product removal and further genetic modification of overexpressing RuBisCo (*rbcL, rbcS*), the titer of isobutyraldehyde produced reached 1.1 g/L in 8 days. In another study, PCC6803 was also modified to express this pathway for isobutanol production.[16] However, the resulting titer of 90 mg/L was lower than that achieved by recombinant PCC7942. To apply *in situ* product removal, oleyl alcohol trap was used. The resulting cultivation achieved an isobutanol production of 240 mg/L. Together, these studies show the importance of having *in situ* product removal.

5.2.1.5. *2-Methyl-1-butanol*

2-Methyl-1-butanol (2MB) is another energy-dense fuel molecule suitable for gasoline substitute. It is also an intermediate for synthesizing other chemicals. 2MB synthesis depends on isoleucine pathway, which provides the precursor to 2MB, 2-keto-3-methylvalerate (Figure 5.2). Similar to isobutanol production pathway, the introduction of a ketoacid decarboxylase (KivD) and an alcohol dehydrogenase (YqhD) would enable the production of 2MB. To engineer photosynthetic 2MB production using cyanobacteria, the citramalate pathway (Figure 5.2. CimA, LeuBCD) was coexpressed with KivD and YqhD.[17] The citramalate pathway enables the synthesis of 2-ketobutyrate, which is the precursor for 2-keto-3-methylvalerate. The resulting strain produced 177.5 mg/L with an average total alcohol productivity of 20 mg/L/day. It is worth noting here that 1-propanol was not observed, which is somewhat counterintuitive as KivD can also act on 2-ketobutyrate. This result indicated that the native isoleucine pathway is very active. In particular, it was shown that acetohydroxyacid synthase (AHAS), catalyzing condensation of 2-ketobutyrate and pyruvate, was demonstrated highly active by an *in vitro* enzyme assay.[17]

5.2.2. *Diols and Polyols*

5.2.2.1. *Glycerol*

Glycerol is naturally synthesized by microbes such as *Saccharomyces cerevisiae* as an osmoprotectant. It is synthesized from dihydroxyacetone phosphate (DHAP), a glycolytic intermediate, in two steps through glycerol-3-phosphate (G3P) (Figure 5.3). To engineer a photosynthetic glycerol producer, phosphoglycerol phosphatase 2 (encoded by *gpp2*) from *S. cerevisiae* was introduced into PCC6803. The resulting strain successfully produced glycerol directly from CO_2.[18] Interestingly, it was observed that salt stress alone can stimulate glycerol production in wild-type PCC6803, up to a titer of 0.7 mM in 10 days. By optimizing the salt stress (200 mM NaCl), the extracellular glycerol concentration increased more than 20% in the *gpp2* overexpressing strain. The final titer reported was 14.3 mM (1,316 mg/L) in 17 days.[18] In another work using PCC7942 as the host, glycerol-3-phosphatase (encoded by *gpp1*) was expressed.[19] This recombinant strain produced glycerol with a concentration of 1.17 g/L, similar titer to that produced by engineered PCC6803. These results show that glycerol production by engineered cyanobacteria can reach a relatively high flux. This is potentially due to the availability of the precursor DHAP, which is the net product of the dark reaction.

5.2.2.2. *2,3-Butanediol*

2,3-Butanediol (2,3-BDO) is a natural fermentation product from some *Bacillus* and *Klebsiella* species. It was discovered as early as 1911 in *Klebsiella oxytoca*.[79] The first step of the 2,3-BDO production pathway is the same as the isobutanol biosynthetic pathway described in Section 5.2.1.4. Therefore, the same driving force for 2,3-butanediol production can be expected. 2,3-BDO is currently one of chemicals produced by engineered cyanobacteria with the highest flux. One of the potential reasons for its high flux, in addition to the strong driving force of the pathway, is the relative non-toxic nature of 2,3-BDO to cyanobacteria.[22] Using the same acetolactate synthase (encoded by *alsS*) from the isobutanol production pathway, photosynthetic 2,3-BDO was successfully produced by engineered PCC6803[21] and PCC7942[22] with titers of 0.585 and 2.38 g/L, respectively. To further improve the efficiency of this pathway, modulation of 5'-untranslated region (5'UTR) was used to balance the gene expression.[80] It was shown that the RBS calculator[81] was capable of introducing RBS with different strengths in cyanobacteria. However, it was observed that the genes positioned further away from

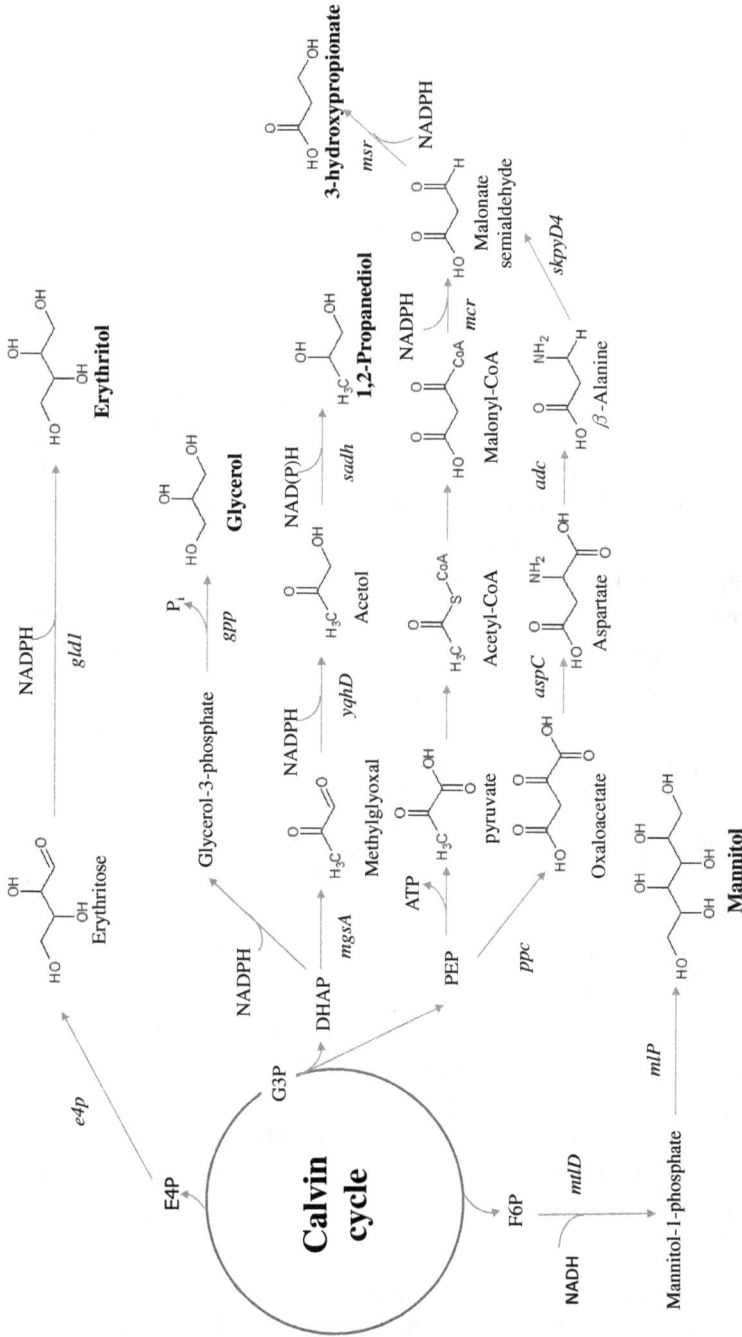

Figure 5.3. Schematics of related pathways originating from CBB cycle intermediates. Abbreviations: *e4p*, erythrose-4-phosphatas; *gld1*, erythrose reductase; *gpp*, glycerol-3-phosphatase; *mgsA*, methylglyoxal synthase; *yqhD*, alcohol dehydrogenase; *sadh*, secondary alcohol dehydrogenase; *acc*, acetyl-CoA carboxylase; *mcr*, malonyl-CoA reductase; *msr*, malonate semialdehyde reductase; *ppc*, phosphoenolpyruvate carboxylase; *aspC*, aspartate aminotransferase; *adc*, aspartate decarboxylase; *skpyD4*, β-alanine aminotransferase; *mtlD*, mannitol-1-phosphate dehydrogenase; *mlP*, mannitol-1-phosphatase.

the promoter were less responsive to the RBS prediction. Furthermore, the 2,3-BDO-producing strain of recombinant PCC7942 was engineered to consume sugars such as glucose or xylose to enable continuous production during diurnal cycles. The resulting production platform enabled 2,3-butanediol production up to 3 g/L in 10 days.[82]

5.2.2.3. *1,2-Propanediol and 1,3-propanediol*

1,2-Propanediol (1,2-PDO) is naturally produced by *Thermoanaerobacterium thermosaccharolyticum* under anaerobic conditions.[83] As a diol, 1,2-propanediol is a useful monomer for producing plastics. By expressing the genes coding for methylglyoxal synthase (*mgsA*), aldehyde reductase (*yqhD*), and glycerol dehydrogenase (*gldA*), recombinant PCC7942 produced 22 mg/L of 1,2-PDO.[23] Because 1,2-propanediol is naturally a fermentative product, the dehydrogenases used were typically NAD-dependent enzymes. However, cyanobacteria use NADPH as its primary electron carrier. Therefore, NADPH-dependent dehydrogenases were recruited, leading to a significant increase of 1,2-PDO produced. The final strain was able to produce 150 mg/L of 1,2-PDO.[23]

1,3-Propanediol (1,3-PDO) is also a natural fermentation product, typically produced by *Klebsiella* species. The natural production pathway for 1,3-PDO is dependent on glycerol. Glycerol is first dehydrated to 3-hydroxypropionaldehyde, which is subsequently reduced to 1,3-PDO. There are two types of the key enzyme glycerol dehydratase. One type of this enzyme is vitamin B12 dependent, while the other is oxygen sensitive. Both of these enzymes were considered to be difficult for expression in cyanobacteria. In addition, the vitamin B12-dependent glycerol dehydratase also requires a reactivase. Interestingly, upon expression of the genes coding for glycerol dehydratase and its reactivase together with NADPH-dependent alcohol dehydrogenase YqhD, photosynthetic production 1,3-PDO has been demonstrated by recombinant PCC7942 with a titer up to 288 mg/L.[20] This result indicated that PCC7942 is capable of producing a sufficient amount of vitamin B12 to adequately support the function of glycerol dehydratase. Further work was done by optimizing the promoters and using higher IPTG and phosphate level.[24] This achieved the titer of 1,3-PDO of 1.22 g/L in 20 days.

5.2.2.4. *Erythritol and mannitol*

Erythritol is used as a sweetener in food and beverage industry because of its non-caloric and non-cariogenic properties. It can be produced through the dephosphorylation of erythrose-4phosphate (E4P), an intermediate of CBB cycle, followed by a reduction to erythritol (Figure 5.3). Recently, this pathway was introduced into PCC6803. The resulting strain achieved the first photosynthetic erythritol production in genetically engineered microorganisms.[25] By expressing a phosphatase (*tm1254*) and an erythrose reductase (*gld1*) from *Hypocrea jecorina* under P_{cpcBA} and P_{Trc} promoters, respectively, on a plasmid with phosphoketolase (*pkt7942*) under P_{psbA2} promoter inserted into the chromosome, this strain was able to produce erythritol up to a titer of 256 mg/L. The overexpressed phosphoketolase was used to enhance the carbon flux from fructose-6-phosphate (F6P) to E4P. Interestingly, as their observation, erythritol forms only when culture is close to stationary phase. Erythritol shows no intracellular accumulation and low extracellular toxicity, suggesting that it may be a good target for photosynthetic production if additional flux can be directed to E4P biosynthesis.

Mannitol is another sweetener also commonly used in the food industry. Using marine cyanobacterium PCC7002 as the host organism, codon-optimized genes for mannitol-1-phosphate dehydrogenase (*mtlD*) and mannitol-1-phosphatase (*mlp*) were overexpressed under the control of P_{psbA2} promoter. These gene overexpressions directed the carbon from F6P to mannitol[26] (Figure 5.3). Together with the knockouts of genes coding for glycogen synthesis, the resulting strain produced mannitol with a concentration of 1.1 g/L in 12 days. Similar to what has been observed in erythritol production, intracellular mannitol concentration was very low in this mannitol-producing strain.

5.2.3. *Fatty Acids and Hydrocarbons*

5.2.3.1. *Fatty acid*

Naturally, triacylglycerides (TAG) are oil bodies that can be extracted and used to produce biodiesel. TAG is naturally produced by oleaginous algae under stress conditions.[84] However, cells do not grow well under stressed conditions, and as a result, direct production and secretion of free fatty acids are potentially more favorable. It was shown that by knocking out fatty-acyl-ACP synthetase, PCC6803 was able to produce 17.5 mg/L of fatty acids in 10 days.[28] With further expression of codon-optimized TesA', an *E. coli* thioesterase without N-terminus periplasmic directing residues, the resulting strain produced 45 mg/L of fatty acids.[29] Subsequently, with combined multiple strategies including expression of codon-optimized TesA' and plant thioesterase, deletion of genes for fatty acid activation, polyhydroxybutyrate, hemolysin-like surface layer protein, cyanophycin, and phosphotransacetylase, engineered PCC 6803 successfully increase fatty acid production from 1.8 to 197 mg/L.[27] Also, by placing lipolytic enzyme under the control of promoters inducible by CO_2 limitation, degradation of membrane diacylglycerol leads to the release of fatty acid,[85] increasing the overall productivity of fatty acids. Fatty acid production has also been demonstrated using recombinant strain of PCC7002. The overexpression of thioesterase (*tesA'*) and RuBisCo (*rbcLS*) while long-chain-fatty-acid-CoA ligase (*fadD*) was knocked out, enabling the strain to produce fatty acids with a titer of 131 mg/L.[31] The same study mentioned that PCC7002 has a higher tolerance to extracellular fatty acid compared with PCC6803 and PCC7942.

5.2.3.2. *Hydrocarbons and fatty alcohols*

Many cyanobacteria naturally produce hydrocarbons from fatty acids using two different pathways.[86] The first and most commonly observed pathway activates fatty acids into fatty-acyl-CoAs, which are reduced and deformylated by fatty-acyl-ACP reductase and aldehyde-deformylating oxygenase, respectively, into alkanes or alkenes (Figure 5.2). The other pathway involves polyketide synthase, which first elongates the carbon chain followed by a decarboxylation to produce hydrocarbons.[86] Interestingly, the two pathways are not present in the same organism, indicating an evolutionary divergence of hydrocarbons biogenesis in cyanobacteria.[86] The level of hydrocarbons naturally produced by cyanobacteria is relatively low compared to the level observed of other biochemical targets.

Photosynthetic fatty alcohol production was first reported in engineered PCC6803 with a titer of 0.2 mg/L through overexpressing *far* gene (fatty-acyl-CoA reductase) from jojoba.[33] Further improvement was made by overexpressing a different fatty-acyl-CoA reductase, encoded by *maqu_2220*. This fatty acyl-CoA reductase is a bifunctional enzyme that reduces both fatty-acyl-CoA and acyl-ACP directly to fatty alcohol.[34] The host was also modified with knockouts of fatty-acyl-ACP reductase (sll0209) and aldehyde-deformylating oxygenase (sll0208), reducing carbon flux to hydrocarbons. The resulting strain produced 2.1 mg/L of fatty alcohols.[34] Compared to the level of fatty acids that engineered cyanobacteria can produce (see Section 5.2.3.1), limitation of fatty alcohol production is likely on either the activation of fatty acids or the reduction of fatty-acyl-ACP or acyl-CoA.

5.2.4. *Organic Acids*

5.2.4.1. *Lactate*

Lactate production by fermentation has been long practiced for its broad usages in the food industry. In addition to application in the food industry, lactic acid can be also used as the monomer of biodegradable polylactic acid plastics, which has been commercialized since the early 21st century. As a means to increase the efficiency of lactate production using solar energy, cyanobacteria have been engineered to produce and secrete lactate. Lactate is naturally produced by one-step reduction from pyruvate. Depending on the particular lactate dehydrogenase used, the resulting lactate could be either D- or L-lactate. Both stereoisomers of lactate have been demonstrated for photosynthetic production by cyanobacteria with varying degrees of productivity (Table 5.1). To engineer a photosynthetic producer for lactate, cyanobacteria have been modified with the overexpression of lactate dehydrogenase,[35,37,87] lactate transporters, and transhydrogenases,[36,40] and genetic knock-down of competing pathways such as phosphoenolpyruvate carboxylase.[38] Heterologous transhydrogenase has also been expressed for lactate biosynthesis because most commonly found lactate dehydrogenases are NAD dependent. As cyanobacteria naturally produce NADPH as the primary electron carrier, transhydrogenase facilitates the transfer of NADPH to NADH, thereby increasing the electron availability for lactate production. Alternatively, NAD-dependent lactate dehydrogenase has also been developed to increase utilization of NADPH.[87] In the various works done on lactate, the highest titer produced by recombinant cyanobacteria reported was 1.8 g/L in 40 days. This relatively high flux product again shows the availability of central glycolytic intermediate for non-native chemical production.

5.2.4.2. *3-Hydroxypropionate*

3-Hydroxypropionate (3HP) is one of the top 12 most value-added platform chemicals identified by the U.S. Department of Energy and is a precursor for various downstream chemicals and polymer synthesis.[88] 3HP is a natural product and can be produced from glycerol. Although cyanobacteria do not naturally produce glycerol, the metabolic pathways for glycerol production have been engineered into cyanobacteria (see Section 5.2.2.1). Using this glycerol-producing strain as a basis, genes coding for glycerol dehydratase and aldehyde reductase were overexpressed. The resulting strain produced 31.7 mg/L of 3HP in 10 days.[19] However, this production was only made possible through dark and anaerobic incubation, indicating potential oxygen sensitivity. In addition to the glycerol-based pathway, 3HP is a natural metabolite found in the 3HP/4-hydroxybutyrate (4HB) cycle for CO_2 fixation. Using the enzyme from 3HP/4HB cycle (Figure 5.3.), both PCC7942 and 6803 have been engineered for photosynthetic 3HP production, resulting in the production of 3HP with titer of 665 and 837 mg/L, respectively.[41,42] A third and synthetic pathway was also constructed in PCC7942. This synthetic pathway utilizes β-alanine biosynthesis. β-Alanine can be converted to malonate semialdehyde, which subsequently is converted into 3HP. Using this pathway, engineered PCC 7942 produced 186 mg/L of 3HP.[41] Both the malonyl-CoA reduction and β-alanine routes were not oxygen sensitive. As a result, these two pathways outperform the glycerol-based pathway and again reinforce the importance of oxygen-tolerant pathways for cyanobacterial engineering.

5.2.4.3. *3-Hydroxybutyrate*

3-Hydroxybutyrate (3HB) is the building block of poly-3-hydroxybutyrate (PHB), a biodegradable plastic that is naturally produced by some microbes, including *Ralstonia eutropha* and PCC6803 as a carbon storage. The first two reactions of PHB biosynthesis are similar to that of 1-butanol biosynthesis (Figure 5.2). This pathway starts by acetyl-CoA condensation to form acetoacetyl-CoA, which is subsequently reduced to 3-hydroxybutyryl-CoA. 3-Hydroxybutyryl-CoA is used as addition monomer to extend the growing chain of PHB. While PHB is a natural product, copolymers of 3-hydroxybutyrate and other hydroxyl acids offer greater flexibility for diverse applications of bioplastics. Thus, the bio-production of 3HB as a monomer has attracted attention recently.[89–91] To produce 3HB from CO_2, PCC6803 was selected as a model host due to its natural ability for PHB accumulation. PCC6803 was modified by deleting PHB synthase gene *phaEC* and expressing heterologous *phaAB* genes from *Ralstonia eutropha* H16 and *tesB* gene from *E. coli*.[91] *phaA* and *phaB* code for the enzymes catalyzing the first two steps of 3HB production pathway. *tesB* codes for a thioesterase. The resulting strain produced 533 mg/L of 3HB in 21 days. It has shown that 3HB productivity from the recombinant strain started out low. After cell growth entered the stationary phase, the productivity of 3HB rapidly increased. This result is likely due to depletion of nitrogen in the medium. As a response, PCC6803 directs carbon flux to acetyl-CoA, facilitating 3HB biosynthesis.

5.2.4.4. *Succinic acid*

Succinic acid is one of the top 12 value-added building block chemicals from biomass suggested by the U.S. Department of Energy.[88] It is currently used as food additive, surfactant, and pharmaceuticals.[92] In addition, it can be converted to other high-volume industrial chemicals such as 1,4-butanediol, tetrahydrofuran, malic acid, and itaconic acid. Until recently, cyanobacteria have been previously thought to lack TCA cycle due to the absence of α-ketoglutarate dehydrogenase. Therefore, natural cyanobacterial secretion of succinic acid results from the reductive branch of TCA cycle and is observed under dark anoxic conditions.[93] Engineered succinic acid-producing *Synechocystis* PCC6803 was constructed by the overexpression of *sigE*, an RNA polymerase sigma factor, and deletion of *ackA*, acetate kinase.[44] The resulting strain produced succinic acid with titer of 113 mg/L under dark anaerobic condition. A subsequent study identified the rate-limiting step as phosphoenolpyruvate carboxylase (Ppc), responsible for converting PEP to oxaloacetate. With *ppc* over expressed, the engineered *Synechocystis* strain improved its succinic acid production titer to 194 mg/L.[45] Additional studies using CRISPR-Cas9[46] and CRISPRi[94] in PCC7942 have been demonstrated for photoautotrophic succinate production. These works also employed a similar dark treatment with nitrogen-depleting conditions. However, the resulting titers of around 0.5 mg/L were significantly less than what has been demonstrated in engineering PCC6803.

Direct photosynthetic conversion of CO_2 to succinate was achieved by the expression of a α-ketoglutarate decarboxylase and succinate semialdehyde dehydrogenase. These two enzymes allow the conversion of α-ketoglutarate to succinic acid. This pathway was identified in *Synechococcus* sp. PCC7002, which changed the broken TCA cycle concept for cyanobacteria.[95] Using this pathway in PCC7942, an engineered strain was able to produce succinate up to 120 mg/L.[47] However, it was noticed that cell growth was severely inhibited upon the expression of this succinate-producing pathway. This growth retardation was rescued by an additional overexpression of PEP carboxylase and citrate synthase, presumably increasing anaplerotic flux to overcome the excessive drain of α-ketoglutarate flux. The resulting strain produced 430 mg/L of succinate in 8 days, presenting a significant improvement.

5.2.4.5. *p-Coumaric acid*

p-Coumaric acid is a precursor for many phenylpropanoids that are potentially beneficial to human health because of their bioactivities such as antiinflammatory and anticancer properties. *p*-Coumaric acid can be produced through the deamination of tyrosine or hydrolysis of cinnamic acid. Recently, PCC6803 was engineered to express tyrosine ammonia lyase (Tal) for *p*-coumaric acid production.[48] This work first reported that no *p*-coumaric acid was produced upon expression of Tal alone. With further investigation, it was discovered that a native gene *slr1573* codes for a laccase that degrades *p*-coumaric acid. Upon knocking out *slr1573* gene and expressing heterologous Tal, the engineered strain produced *p*-coumaric acid with a titer of 82.6 mg/L. This same group further engineered this *p*-coumaric acid producing strain for *p*-coumarate-3-hydroxylase expression. The resulting strain produced caffeic acid with a titer of 7.2 mg/L.[49]

5.2.5. *Olefins*

5.2.5.1. *Ethylene*

Olefins are frequently used as monomers for chain-growth polymerization in the production of plastics. Ethylene, the simplest olefin, is the monomer for producing polyethylene, a common and broadly used plastic in commercial products such as plastic bags. Biologically, ethylene is a natural product and serves as a hormone in plants. Using this pathway, PCC7942 was used as a model host expressing 1-aminocyclopropane-1-carboxylate synthase (ACS) and aminocyclopropane carboxylate oxidase (ACO). Through the expression of these two enzymes, ethylene production from *S*-adenosylmethionine (SAM) was demonstrated with a productivity of 81.6 nL/mL/day/OD.[54] However, as the plant pathway for ethylene production is more complex, it is less investigated for use in microbial conversion of ethylene. On the other hand, species of *Pseudomonas* are also known to produce ethylene through using a single enzyme called ethylene-forming enzyme (Efe). Efe catalyzes the conversion of α-ketoglutarate to ethylene. Due to the simplicity of this ethylene production pathway, it is more frequently used. Photosynthetic ethylene production have been demonstrated as early as in the 1990s.[96,97] However, the production of ethylene using engineered cyanobacteria expressing Efe was reported with low productivity (Table 5.1) due to instability of the plasmid used. Further studies revealed that mutations occur on *efe* gene sequence after several generations of cell growth, which ultimately stopped ethylene production.[51,52] To improve the stability and productivity of ethylene biosynthesis in cyanobacteria, the predicted "hot spot" regions for mutation in the coding sequence were modified.[52] The *efe* gene was also codon optimized for expression in cyanobacteria. The resulting engineered strain of PCC6803 was able to achieve significant level of ethylene production with a peak productivity of 171 mg/L/day,[52] representing one of the top productivities demonstrated for cyanobacterial chemical production.

5.2.5.2. *Isoprene*

Isoprenoid biosynthesis in most prokaryotes depends on the methyleryth-ritol-4-phosphate (MEP) pathway (Figure 5.4). On the other hand, archaea and eukaryotes typically use the mevalonate (MVA) pathway. Both the MEP and MVA pathways end up synthesizing isopentenyl pyrophosphate (IPP) and dimethylallyl diphosphate (DMAPP), which is subsequently used for downstream terpenoid biosynthesis. DMAPP, in particular, is the direct substrate for isoprene synthase (IspS) in isoprene biosynthesis. Isoprene synthase is an enzyme naturally found in plants, especially in polar and kudzu vine. In 2010, the first photosynthetic isoprene production was demonstrated by engineered PCC6803 expressing IspS expression.[56] However, the resulting titer of 50 µg/g dry cell weight was relatively low. To improve the production of isoprene, a gaseous/aqueous two-phase photobioreactor was used, resulting in a titer of 0.35 mg/L in 8 days.[57] Similar to isobutyraldehyde (Section 5.2.1.4) production,[16] this increase of isoprene titer shows that *in situ* product removal is a useful method to improve the production of volatile compounds. Furthermore, heterologous coexpression of mevalonate pathway with the aim of increasing the carbon flux into DMAPP biosynthesis also increased the isoprene production when compared to relying on only the MEP pathway.[58] However, the resulting titers remained relatively low. Recently, a sharp increase to the titer (up to 1.26 g/L) of isoprene production was made by increasing DMAPP to IPP ratio, constructing protein fusions for reducing the need for substrate diffusion, conducting long-term production, and overexpressing the bottleneck gene *ispG*, coding for HMB-PP synthase.[59]

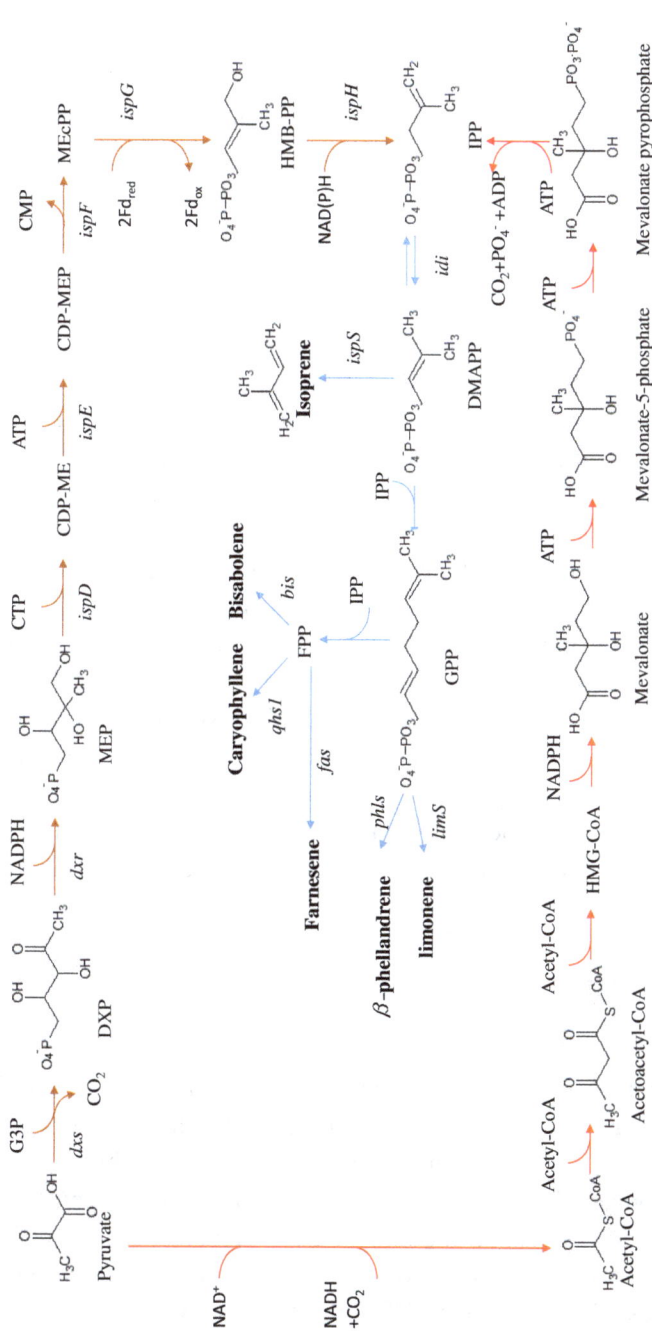

Figure 5.4. Schematics of isoprenoids and terpenoids biosynthesis based on MEP and MVA pathways. The genes *dxs*, *dxr*, *ispD*, *ispE*, *ispF*, *ispG*, and *ispH* are related in MEP pathway (brown arrows). The red arrows indicate the reaction of MVA pathway. Abbreviations: *idi*, IPP isomerase; *ispS*, isoprene synthase; *limS*, limonene synthase; *phls*, β-phellandrene synthase; *fas*, farnesene synthase; *qhs1*, β-caryophyllene synthase; *bis*, bisabolene synthase.

5.2.5.3. Terpenoids

Terpenoid is a large class of biochemicals extended from DMAPP and IPP. The five-carbon DMAPP and IPP can condense and form geranyl-pyrophosphate (GPP), which is a precursor for monoterpenes. GPP can further react with another molecule of IPP to form farnesyl pyrophosphate (FPP), which is the precursor for sesquiterpenes. Thus far, the production of some monoterpenoids and sesquiterpenoids using engineering cyano-bacteria have been demonstrated. Monoterpenes such as β-phellandrene and L-limonene are produced through the expression of β-phellandrene synthase and L-limonene synthase, respectively. As indicated in Table 5.1, most studies aiming to produce these monoterpenes primarily focus on enhancing carbon flux for monoterpene synthesis through the overex-pression of upstream MEP pathway[60] and the knockout of glycogen biosynthesis,[61] optimizing enzyme expression through codon optimiza-tion[64] and protein fusions,[65] and conducting production with *in situ* prod-uct removal.[61,62] Currently, the highest titer of limonene and β-phellandrene achieved is around 4 and 0.9 mg/L, respectively. Considering the produc-tion of isoprene has been demonstrated in the g/L level, it is likely that the bottleneck for terpenoid production is downstream of IPP and DMAPP biosynthesis. Going beyond monoterpenes, sesquiterpenes can be synthesized upon the expression of suitable enzymes to convert FPP into the target sesquiterpenes. As such, cyanobacteria have been engi-neered to produce β-caryophyllene,[67] farnesene,[68] and bisabolene.[61] However, the current productivity for these molecules remains relatively low (Table 5.1).

5.2.6. Sugars

As cyanobacteria are typically less tolerant to the non-native biochemicals that they are programmed to produce than heterotrophs, cyanobacteria have been proposed to produce mono- or di-saccharides, which can be subsequently converted to target biochemicals by heterotrophs with ease. Sugars are generally non-toxic. Naturally, many cyanobacteria are capable of producing sucrose as an osmoprotectant with intracellular concentra-tion reaching as high as 300 mM,[98] indicating the vast potential for cyanobacterial sugar production. The first step to constructing a photosynthetic sugar producer is to engineer cyanobacteria for sucrose secretion. CscB, a symporter for protons and sucrose, was expressed in PCC7942.[70] This resulting strain was further engineered by inhibition of invertase, the enzyme responsible for hydrolyzing sucrose and glycogen synthesis. The final titer achieved was around 2.7 g/L in 3 days. Alternatively, if fructose and glucose are the desired products, invertase could be overexpressed instead of getting inhibited. This approach has been done to PCC7942 for the production of these monosaccharides.[35]

Similar engineering methods as those used in PCC7942 were applied to the recently isolated UTEX 2973.[72] Gene *cscB* under IPTG-inducible promoter P_{lac} was introduced into the genome of UTEX 2973. Subsequently using salt stress, the recombinant strain produced up to 3.34 g/L of sucrose in 94 h. This study then analyzed the ability of a culture to be reused for sequential rounds of sucrose production. It was observed that the cells were able to continue sucrose production up to seven times, yielding a total cumulative titer of 8.7 g/L sucrose. These results showed that this engineered strain was stable for the direct photosynthetic sugar production. Compared to most other engineered chemical products, these studies on recombinant photosynthetic sugars productions typically resulted in relatively high titers in cyanobacteria, indicating that the natural flux in sugar phosphates is expectedly high in cyanobacteria metabolism.

5.3. Conclusion

Using metabolically engineered cyanobacteria for the production of chemicals is advantageous over conventional sugar-based fermentation because it avoids the need for repeated growth and deconstruction of plant biomass. As discussed in this chapter, many of the major biochemicals of interest have been demonstrated for the production from CO_2 by cyanobacteria, indicating the feasibility of photosynthetic CO_2-based chemical production. However, the current limitation for this technology to be vastly commercialized is its relative low titer and productivity compared to conventional fermentations. Currently, the best production titers demonstrated in the literature is in the grams per liter range. These chemicals include ethanol, 2,3-butanediol, lactate, glycerol, isobutyraldehyde, and sugars. These chemicals are generally non-toxic or are easily removed from production culture, which also reduces any potential toxic effects. Furthermore, most of these chemicals are derived from glycolytic metabolites such as pyruvate. As reported recently,[59] pyruvate pool is more than 500-fold larger than acetyl-CoA pool in strain PCC7942. Therefore, it is not surprising that the chemicals with higher productivities are those derived from pyruvate. While the current productivities and titers of photosynthetic biochemical productions are still relatively low compared to that of conventional sugar-based fermentation, the use of photosynthetic microbes does not require arable land, can be used with waste water, and can directly use flue gas produced by power, steel, and other manufacture industries. All of these characteristics outperform the conventional plant-based bio-productions. With further improvements in protein expression, genetic control, photosynthetic rate, and production process design, photosynthetic microbes such as cyanobacteria can be expected to bring revolutionizing technology for achieving sustainability.

References

1. Kaneko, T. *et al.* Sequence analysis of the genome of the unicellular cyanobacterium *Synechocystis* sp. strain PCC6803. II. Sequence determination of the entire genome and assignment of potential protein-coding regions. *DNA Res.* **3**, 109–136 (1996).
2. Shestakov, S., & Khyen, N.T. Evidence for genetic transformation in blue-green alga *Anacystis nidulans*. *Mol. Gen. Genetics MGG* **107**, 372–375 (1970).
3. Batterton Jr, J.C., & Van Baalen, C. Growth responses of blue-green algae to sodium chloride concentration. *Arch. Mikrobiol.* **76**, 151–165 (1971).
4. Nomura, C.T., Sakamoto, T., & Bryant, D.A. Roles for heme–copper oxidases in extreme high-light and oxidative stress response in the cyanobacterium *Synechococcus* sp. PCC 7002. *Arch. Mikrobiol.* **185**, 471–479 (2006).
5. Yu, J.J. *et al.* Synechococcus elongatus UTEX 2973, a fast growing cyanobacterial chassis for biosynthesis using light and CO_2. *Sci. Rep.* **5**, 8132 (2015).

6. Deng, M.-D., & Coleman, J.R. Ethanol synthesis by genetic engineering in cyanobacteria. *Appl. Environ. Microbiol.* **65**, 523–528 (1999).

7. Dexter, J., & Fu, P. Metabolic engineering of cyanobacteria for ethanol production. *Energy Environ. Sci.* **2**, 857–864 (2009).

8. Gao, Z., Zhao, H., Li, Z., Tan, X., & Lu, X. Photosynthetic production of ethanol from carbon dioxide in genetically engineered cyanobacteria. *Energy Environ. Sci.* **5**, 9857–9865 (2012).

9. Lan, E.I., Ro, S.Y., & Liao, J.C. Oxygen-tolerant coenzyme A-acylating aldehyde dehydrogenase facilitates efficient photosynthetic n-butanol biosynthesis in cyanobacteria. *Energy Environ. Sci.* **6**, 2672–2681 (2013).

10. Kusakabe, T. *et al.* Engineering a synthetic pathway in cyanobacteria for isopropanol production directly from carbon dioxide and light. *Metab. Eng.* **20**, 101–108 (2013).

11. Hirokawa, Y., Suzuki, I., & Hanai, T. Optimization of isopropanol production by engineered cyanobacteria with a synthetic metabolic pathway. *J. Biosci. Bioeng.* **119**, 585–590 (2015).

12. Hirokawa, Y., Dempo, Y., Fukusaki, E., & Hanai, T. Metabolic engineering for isopropanol production by an engineered cyanobacterium, *Synechococcus elongatus* PCC 7942, under photosynthetic conditions. *J. Biosci. Bioeng.* **123**, 39–45 (2017).

13. Lan, E.I., & Liao, J.C. Metabolic engineering of cyanobacteria for 1-butanol production from carbon dioxide. *Metab. Eng.* **13**, 353–363 (2011).

14. Lan, E.I., & Liao, J.C. ATP drives direct photosynthetic production of 1-butanol in cyanobacteria. *Proc. Natl. Acad. Sci.* **109**, 6018–6023 (2012).

15. Atsumi, S., Higashide, W., & Liao, J.C. Direct photosynthetic recycling of carbon dioxide to isobutyraldehyde. *Nat. Biotechnol.* **27**, 1177–1180 (2009).

16. Varman, A.M., Xiao, Y., Pakrasi, H.B., & Tang, Y.J. Metabolic engineering of *Synechocystis* sp. strain PCC 6803 for isobutanol production. *Appl. Environ. Microbiol.* **79**, 908–914 (2013).

17. Shen, C.R., & Liao, J.C. Photosynthetic production of 2-methyl-1-butanol from CO_2 in cyanobacterium *Synechococcus elongatus* PCC7942 and characterization of the native acetohydroxyacid synthase. *Energy Environ. Sci.* **5**, 9574–9583 (2012).

18. Savakis, P. *et al.* Photosynthetic production of glycerol by a recombinant cyanobacterium. *J. Biotechnol.* **195**, 46–51 (2015).

19. Wang, Y., Tao, F., Ni, J., Li, C., & Xu, P. Production of C3 platform chemicals from CO_2 by genetically engineered cyanobacteria. *Green Chem.* **17**, 3100–3110 (2015).

20. Hirokawa, Y., Maki, Y., Tatsuke, T., & Hanai, T. Cyanobacterial production of 1,3-propanediol directly from carbon dioxide using a synthetic metabolic pathway. *Metab. Eng.* **34**, 97–103 (2016).

21. Savakis, P.E., Angermayr, S.A., & Hellingwerf, K.J. Synthesis of 2,3-butanediol by *Synechocystis* sp. PCC6803 via heterologous expression of a catabolic pathway from lactic acid-and enterobacteria. *Metab. Eng.* **20**, 121–130 (2013).

22. Oliver, J.W., Machado, I.M., Yoneda, H., & Atsumi, S. Cyanobacterial conversion of carbon dioxide to 2,3-butanediol. *Proc. Natl. Acad. Sci.* **110**, 1249–1254 (2013).

23. Li, H., & Liao, J.C. Engineering a cyanobacterium as the catalyst for the photosynthetic conversion of CO_2 to 1,2-propanediol. *Microb. Cell Fact.* **12**, 1 (2013).

24. Hirokawa, Y., Maki, Y., & Hanai, T. Improvement of 1,3-propanediol production using an engineered cyanobacterium, *Synechococcus elongatus* by optimization of the gene expression level of a synthetic metabolic pathway and production conditions. *Metab. Eng.* **39**, 192–199 (2017).

25. van der Woude, A.D. *et al.* Genetic engineering of *Synechocystis* PCC6803 for the photoautotrophic production of the sweetener erythritol. *Microb. Cell Fact.* **15**, 1 (2016).

26. Jacobsen, J.H., & Frigaard, N.-U. Engineering of photosynthetic mannitol biosynthesis from CO_2 in a cyanobacterium. *Metab. Eng.* **21**, 60–70 (2014).

27. Liu, X., Sheng, J., & Curtiss III, R. Fatty acid production in genetically modified cyanobacteria. *Proc. Natl. Acad. Sci.* **108**, 6899–6904 (2011).

28. Gao, Q., Wang, W., Zhao, H., & Lu, X. Effects of fatty acid activation on photosynthetic production of fatty acid-based biofuels in *Synechocystis* sp. PCC6803. *Biotechnol. Biofuels* **5**, 1 (2012).

29. Ruffing, A.M., & Jones, H.D. Physiological effects of free fatty acid production in genetically engineered *Synechococcus elongatus* PCC 7942. *Biotechnol. Bioeng.* **109**, 2190–2199 (2012).

30. Ruffing, A.M. Borrowing genes from *Chlamydomonas reinhardtii* for free fatty acid production in engineered cyanobacteria. *J. Appl. Psychol.* **25**, 1495–1507 (2013).

31. Ruffing, A.M. Improved free fatty acid production in cyanobacteria with *Synechococcus* sp. PCC 7002 as host. *Front. Bioeng. Biotechnol.* **2**, 17 (2014).

32. Lee, H.J. *et al.* Photosynthetic CO_2 conversion to fatty acid ethyl esters (FAEEs) using engineered cyanobacteria. *J. Agric. Food Chem.* **65**(6), 1087–1092 (2017).

33. Tan, X. *et al.* Photosynthesis driven conversion of carbon dioxide to fatty alcohols and hydrocarbons in cyanobacteria. *Metab. Eng.* **13**, 169–176 (2011).
34. Yao, L., Qi, F., Tan, X., & Lu, X. Improved production of fatty alcohols in cyanobacteria by metabolic engineering. *Biotechnol. Biofuels* **7**, 1 (2014).
35. Niederholtmeyer, H., Wolfstädter, B.T., Savage, D.F., Silver, P.A., & Way, J.C. Engineering cyanobacteria to synthesize and export hydrophilic products. *Appl. Environ. Microbiol.* **76**, 3462–3466 (2010).
36. Varman, A.M., Yu, Y., You, L., & Tang, Y.J. Photoautotrophic production of D-lactic acid in an engineered cyanobacterium. *Microb. Cell Fact.* **12**, 1 (2013).
37. Joseph, A. *et al.* Utilization of lactic acid bacterial genes in *Synechocystis* sp. PCC 6803 in the production of lactic acid. *Biosci. Biotechnol. Biochem.* **77**, 966–970 (2013).
38. Angermayr, S.A. *et al.* Exploring metabolic engineering design principles for the photosynthetic production of lactic acid by *Synechocystis* sp. PCC6803. *Biotechnol. Biofuels* **7**, 1 (2014).
39. Angermayr, S.A., & Hellingwerf, K.J. On the use of metabolic control analysis in the optimization of cyanobacterial biosolar cell factories. *J. Phys. Chem. B* **117**, 11169–11175 (2013).
40. Angermayr, S.A., Paszota, M., & Hellingwerf, K.J. Engineering a cyanobacterial cell factory for production of lactic acid. *Appl. Environ. Microbiol.* **78**, 7098–7106 (2012).
41. Lan, E.I. *et al.* Metabolic engineering of cyanobacteria for photosynthetic 3-hydroxypropionic acid production from CO_2 using *Synechococcus elongatus* PCC 7942. *Metab. Eng.* **31**, 163–170 (2015).
42. Wang, Y. *et al.* Biosynthesis of platform chemical 3-hydroxypropionic acid (3-HP) directly from CO_2 in cyanobacterium *Synechocystis* sp. PCC6803. *Metab. Eng.* **34**, 60–70 (2016).
43. Wang, B., Pugh, S., Nielsen, D.R., Zhang, W., & Meldrum, D.R. Engineering cyanobacteria for photosynthetic production of 3-hydroxybutyrate directly from CO_2. *Metab. Eng.* **16C**, 68–77 (2013).
44. Osanai, T. *et al.* Genetic manipulation of a metabolic enzyme and a transcriptional regulator increasing succinate excretion from unicellular cyanobacterium. *Front. Microbiol.* **6**, 1064 (2015).
45. Hasunuma, T., Matsuda, M., & Kondo, A. Improved sugar-free succinate production by *Synechocystis* sp. PCC6803 following identification of the limiting steps in glycogen catabolism. *Metab. Eng. Commun.* **3**, 130–141 (2016).
46. Li, H. *et al.* CRISPR-Cas9 for the genome engineering of cyanobacteria and succinate production. *Metab. Eng.* **38**, 293–302 (2016).

47. Lan, E.I., & Wei, C.T. Metabolic engineering of cyanobacteria for the photo-synthetic production of succinate. *Metab. Eng.* **38**, 483–493 (2016).

48. Xue, Y., Zhang, Y., Cheng, D., Daddy, S., & He, Q. Genetically engineering *Synechocystis* sp. Pasteur Culture Collection 6803 for the sustainable production of the plant secondary metabolite p-coumaric acid. *Proc. Natl. Acad. Sci.* **111**, 9449–9454 (2014).

49. Xue, Y., Zhang, Y., Grace, S., & He, Q. Functional expression of an *Arabidopsis* p450 enzyme, *p*-coumarate-3-hydroxylase, in the cyanobacterium *Synechocystis* PCC6803 for the biosynthesis of caffeic acid. *J. Appl. Phycol.* **26**, 219–226 (2014).

50. Chin, T., Sano, M., Takahashi, T., Ohara, H., & Aso, Y. Photosynthetic production of itaconic acid in *Synechocystis* sp. PCC6803. *J. Biotechnol.* **195**, 43–45 (2015).

51. Takahama, K., Matsuoka, M., Nagahama, K., & Ogawa, T. Construction and analysis of a recombinant cyanobacterium expressing a chromosomally inserted gene for an ethylene-forming enzyme at the psbAI locus. *J. Biosci. Bioeng.* **95**, 302–305 (2003).

52. Ungerer, J. *et al.* Sustained photosynthetic conversion of CO_2 to ethylene in recombinant cyanobacterium *Synechocystis* 6803. *Energy Environ. Sci.* **5**, 8998–9006 (2012).

53. Xiong, W. *et al.* The plasticity of cyanobacterial metabolism supports direct CO_2 conversion to ethylene. *Nat. Plants* **1**, 15053 (2015).

54. Jindou, S. *et al.* Engineered platform for bioethylene production by a cyano-bacterium expressing a chimeric complex of plant enzymes. *ACS Synth. Biol.* **3**, 487–496 (2014).

55. Guerrero, F., Carbonell, V., Cossu, M., Correddu, D., & Jones, P.R. Ethylene synthesis and regulated expression of recombinant protein in *Synechocystis* sp. PCC 6803. *PLoS One* **7**, e50470 (2012).

56. Lindberg, P., Park, S., & Melis, A. Engineering a platform for photosynthetic isoprene production in cyanobacteria, using *Synechocystis* as the model organism. *Metab. Eng.* **12**, 70–79 (2010).

57. Bentley, F.K., & Melis, A. Diffusion-based process for carbon dioxide uptake and isoprene emission in gaseous/aqueous two-phase photobioreactors by photosynthetic microorganisms. *Biotechnol. Bioeng.* **109**, 100–109 (2012).

58. Bentley, F.K., Zurbriggen, A., & Melis, A. Heterologous expression of the mevalonic acid pathway in cyanobacteria enhances endogenous carbon partitioning to isoprene. *Mol. Plant* **7**, 71–86 (2014).

59. Gao, X. *et al*. Engineering the methylerythritol phosphate pathway in cyanobacteria for photosynthetic isoprene production from CO_2. *Energy Environ. Sci.* **9**, 1400–1411 (2016).

60. Halfmann, C., Gu, L., & Zhou, R. Engineering cyanobacteria for the production of a cyclic hydrocarbon fuel from CO_2 and H_2O. *Green Chem.* **16**, 3175–3185 (2014).

61. Davies, F.K., Work, V.H., Beliaev, A.S., & Posewitz, M.C. Engineering limonene and bisabolene production in type and a glycogen-deficient mutant of *Synechococcus* sp. PCC 7002. *Front Bioeng. Biotechnol.* **2**, 1–11 (2014).

62. Kiyota, H., Okuda, Y., Ito, M., Hirai, M.Y., & Ikeuchi, M. Engineering of cyanobacteria for the photosynthetic production of limonene from CO_2. *J. Biotechnol.* **185**, 1–7 (2014).

63. Wang, X. *et al*. Enhanced limonene production in cyanobacteria reveals photosynthesis limitations. *Proc. Natl. Acad. Sci.* **113**, 14225–14230 (2016).

64. Bentley, F.K., García-Cerdán, J.G., Chen, H.-C., & Melis, A. Paradigm of monoterpene (β-phellandrene) hydrocarbons production via photosynthesis in cyanobacteria. *Bioenergy Res.* **6**, 917–929 (2013).

65. Formighieri, C., & Melis, A. A phycocyanin·phellandrene synthase fusion enhances recombinant protein expression and β-phellandrene (monoterpene) hydrocarbons production in *Synechocystis* (cyanobacteria). *Metab. Eng.* **32**, 116–124 (2015).

66. Formighieri, C., & Melis, A. Regulation of β-phellandrene synthase gene expression, recombinant protein accumulation, and monoterpene hydrocarbons production in *Synechocystis* transformants. *Planta* **240**, 309–324 (2014).

67. Reinsvold, R.E., Jinkerson, R.E., Radakovits, R., Posewitz, M.C., & Basu, C. The production of the sesquiterpene β-caryophyllene in a transgenic strain of the cyanobacterium *Synechocystis*. *J. Plant Physiol.* **168**, 848–852 (2011).

68. Halfmann, C., Gu, L., Gibbons, W., & Zhou, R. Genetically engineering cyanobacteria to convert CO_2, water, and light into the long-chain hydrocarbon farnesene. *Appl. Microbiol. Biotechnol.* **98**, 9869–9877 (2014).

69. Englund, E. *et al*. Production of squalene in *Synechocystis* sp. PCC 6803. *PLoS One* **9**, e90270 (2014).

70. Ducat, D.C., Avelar-Rivas, J.A., Way, J.C., & Silver, P.A. Rerouting carbon flux to enhance photosynthetic productivity. *Appl. Environ. Microbiol.* **78**, 2660–2668 (2012).

71. Du, W., Liang, F., Duan, Y., Tan, X., & Lu, X. Exploring the photosynthetic production capacity of sucrose by cyanobacteria. *Metab. Eng.* **19**, 17–25 (2013).

72. Song, K., Tan, X., Liang, Y., & Lu, X. The potential of *Synechococcus elongatus* UTEX 2973 for sugar feedstock production. *Appl. Microbiol. Biotechnol.* **100**(18), 7865–7875 (2016).

73. Tan, X., Du, W., & Lu, X. Photosynthetic and extracellular production of glucosylglycerol by genetically engineered and gel-encapsulated cyanobacteria. *Appl. Microbiol. Biotechnol.* **99**, 2147–2154 (2015).

74. Zhou, J., Zhang, H., Zhang, Y., Li, Y., & Ma, Y. Designing and creating a modularized synthetic pathway in cyanobacterium *Synechocystis* enables production of acetone from carbon dioxide. *Metab. Eng.* **14**, 394–400 (2012).

75. Lai, M.C., & Lan, E.I. Advances in metabolic engineering of cyanobacteria for photosynthetic biochemical production. *Metabolites* **5**, 636–658 (2015).

76. Atsumi, S. *et al.* Metabolic engineering of *Escherichia coli* for 1-butanol production. *Metab. Eng.* **10**, 305–311 (2008).

77. Steen, E.J. *et al.* Metabolic engineering of *Saccharomyces cerevisiae* for the production of n-butanol. *Microb. Cell Fact.* **7**, 36 (2008).

78. Atsumi, S., Hanai, T., & Liao, J.C. Non-fermentative pathways for synthesis of branched-chain higher alcohols as biofuels. *Nature* **451**, 86–89 (2008).

79. Walpole, G. The Action of Bacillus actis aerogenes on Glucose and Mannitol. Part II.—The Investigation of the 2:3 Butanediol and the Acetylmethylcarbinol Formed; the Effect of Free Oxygen on their Production; the Action of *B. Lactis* aerogenes on Fructose. *Proc. R. Soc. Lond., B, Biol. Sci.* **83**, 272–286 (1911).

80. Oliver, J.W., Machado, I.M., Yoneda, H., & Atsumi, S. Combinatorial optimization of cyanobacterial 2,3-butanediol production. *Metab. Eng.* **22**, 76–82 (2014).

81. Borujeni, A.E., Channarasappa, A.S., & Salis, H.M. Translation rate is controlled by coupled trade-offs between site accessibility, selective RNA unfolding and sliding at upstream standby sites. *Nucleic Acids Res.* **42**, 2646–2659 (2014).

82. Oliver, J.W., Machado, I.M., Yoneda, H., & Atsumi, S. Combinatorial optimization of cyanobacterial 2,3-butanediol production. *Metab. Eng.* **22**, 76–82 (2014).

83. Cameron, D.C., & Cooney, C.L. A Novel Fermentation: The Production of R (-)-1,2-Propanediol and Acetol by *Clostridium thermosaccharolyticum*. *Nat. Biotechnol.* **4**, 651–654 (1986).

84. Hu, Q. *et al.* Microalgal triacylglycerols as feedstocks for biofuel production: Perspectives and advances. *Plant J.* **54**, 621–639 (2008).

85. Liu, X., Fallon, S., Sheng, J., & Curtiss, R. CO_2-limitation-inducible Green Recovery of fatty acids from cyanobacterial biomass. *Proc. Natl. Acad. Sci.* **108**, 6905–6908 (2011).

86. Coates, R.C. *et al.* Characterization of cyanobacterial hydrocarbon composition and distribution of biosynthetic pathways. *PLoS One* **9**, e85140 (2014).

87. Li, C. *et al.* Enhancing the light-driven production of d-lactate by engineering cyanobacterium using a combinational strategy. *Sci. Rep.* **5**, 9777 (2015).

88. Werpy, T. *et al.* (DTIC Document, 2004).

89. Lin, Z. *et al.* Metabolic engineering of *Escherichia coli* for poly (3-hydroxybutyrate) production via threonine bypass. *Microb. Cell Fact.* **14**, 1 (2015).

90. Wei, X.-X., Zheng, W.-T., Hou, X., Liang, J., & Li, Z.-J. Metabolic Engineering of *Escherichia coli* for Poly (3-hydroxybutyrate) Production under Microaerobic Condition. *Biomed Res. Int.* **2015**, 789315 (2015).

91. Wang, B., Pugh, S., Nielsen, D.R., Zhang, W., & Meldrum, D.R. Engineering cyanobacteria for photosynthetic production of 3-hydroxybutyrate directly from CO_2. *Metab. Eng.* **16**, 68–77 (2013).

92. Beauprez, J.J., De Mey, M., & Soetaert, W.K. Microbial succinic acid production: Natural versus metabolic engineered producers. *Process Biochem.* **45**, 1103–1114 (2010).

93. McNeely, K., Xu, Y., Bennette, N., Bryant, D.A., & Dismukes, G.C. Redirecting reductant flux into hydrogen production via metabolic engineering of fermentative carbon metabolism in a cyanobacterium. *Appl. Environ. Microbiol.* **76**, 5032–5038 (2010).

94. Huang, C.H. *et al.* CRISPR interference (CRISPRi) for gene regulation and succinate production in cyanobacterium *S. elongatus* PCC7942. *Microb. Cell Fact.* **15**, 196 (2016).

95. Zhang, S., & Bryant, D.A. The tricarboxylic acid cycle in cyanobacteria. *Science* **334**, 1551–1553 (2011).

96. Sakai, M., Ogawa, T., Matsuoka, M., & Fukuda, H. Photosynthetic conversion of carbon dioxide to ethylene by the recombinant cyanobacterium, *Synechococcus* sp. PCC7942, which harbors a gene for the ethylene-forming enzyme of *Pseudomonas syringae*. *J. Ferment. Bioeng.* **84**, 434–443 (1997).

97. Fukuda, H. *et al.* Heterologous expression of the gene for the ethylene-forming enzyme from *Pseudomonas syringae* in the cyanobacterium *Synechococcus*. *Biotechnol. Lett.* **16**, 1–6 (1994).

98. Klähn, S., & Hagemann, M. Compatible solute biosynthesis in cyanobacteria. *Environ. Microbiol.* **13**, 551–562 (2011).

Chapter 6

Pentose Phosphate Pathway and Its Metabolic Engineering Applications

*Ying Wang and Chun Li**

Department of Biological Engineering,
Institute for Biotransformation and Synthetic Biosystem,
School of Life Science, Beijing Institute of Technology,
Beijing 100081, China
**lichun@bit.edu.cn*

6.1. Introduction

With the development of technology and industry in the past decades, our society has dramatically changed, making life more convenient for human beings. The petroleum-based revolution in energy industry and chemical synthesis contributes greatly to human prosperity.[1] However, the petroleum-based production mode brings disadvantages such as environmental crisis, high energy inputs, and non-sustainable development. Hence, it is urgent to develop alternative ways to fully or partly replace petroleum-based production of fuels and chemicals.

Microbial manufacturing has been developed as an alternative way for chemical synthesis.[2] Microbes have been widely used in the biosynthesis of various chemicals including proteins, amino acids, alcohols, fatty acids, natural products, and so on.[3,4] Microbes possess many characteristics such as simple genetic background, fast growing rate, and mature genetic manipulation tools so that they are ideal hosts for the biosynthesis of compounds and the further industrial production. Modern science and technologies have been providing new approaches for the construction of microbial cell factory for the production of compounds. Among them, metabolic engineering and synthetic biology have been found to be efficient methods for chemical synthesis.[4–6] One of the key points in metabolic engineering is metabolic pathway construction and optimization.

Pentose phosphate pathway (PPP), which is also called the phosphogluconate pathway and hexose monophosphate shunt, is primarily an anabolic pathway branching from glycolysis at the first committed step of glucose metabolism.[7–10] In the 1930s, Otto Warburg and Fritz Lipman discovered glucose phosphate dehydrogenase and 6-phosphogluconate dehydrogenase, directing the metabolic flux of glucose into an unknown pathway except glycolysis. Meanwhile, they found that $NADP^+$ was required for the oxidation of glucose-6-phosphate (G6P), which was proven to be the first step of the PPP.[11] After that, Frank Dickens *et al.* fully elucidated the whole PPP in 1950s based on the previous studies so that the PPP was also called Warburg–Dickens PPP.[10,12]

The PPP is an important pathway alternative and complementary to glycolysis as well as a main producer of NADPH and pentose phosphates. Therefore, the metabolic engineering of the PPP has been widely studied for the production of diverse compounds such as acids, alcohols, proteins, and natural products employing either prokaryotic or eukaryotic microbes.

6.2. Fundamentals of PPP in Biological Systems

The PPP plays an important role in anabolic metabolism, especially the metabolism of carbohydrate.[13] Glucose is catalyzed by hexokinase and converted to G6P, which is a primary substrate for the PPP (Figure 6.1). The PPP generates NADPH and pentoses as well as ribose 5-phosphate, a precursor for the synthesis of nucleotides.[9,14]

For most organisms, the PPP takes place in cytosol while most steps are conducted in plastids in plants.[15] The PPP can be divided into two phases including oxidative phase and non-oxidative phase (Figure 6.1).[7,9]

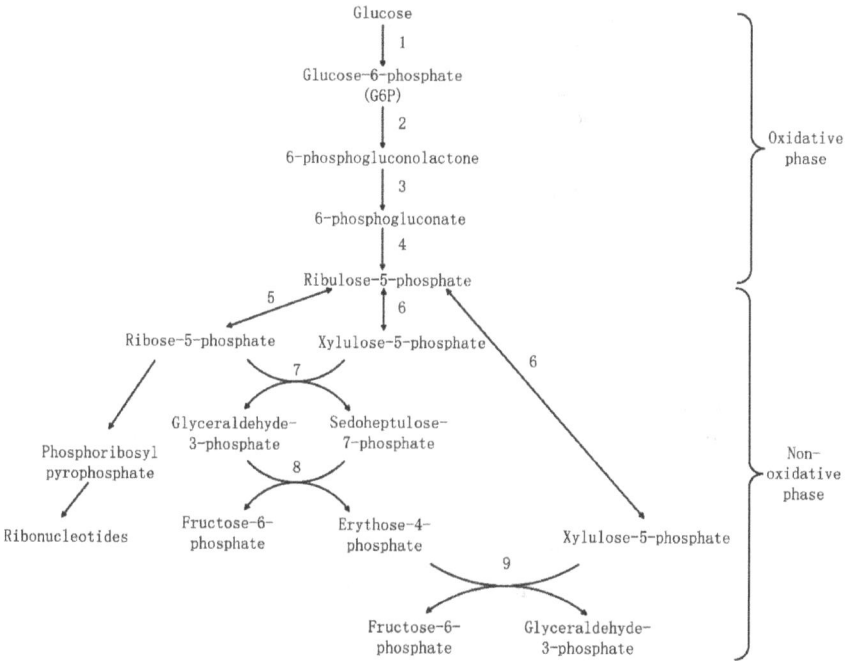

Figure 6.1. Pentose phosphate pathway. Oxidative and non-oxidative phases are involved in the PPP. The PPP begins with the dehydrogenation of G6P. NADPH is the yield in the reactions of either from G6P to 6-phosphogluconolactone or from 6-phosphogluconate to ribulose-5-phosphate (Ru5P) in the oxidative phase of the PPP. No ATP is generated in the PPP. 1: Hexokinases; 2: glucose-6-phosphate dehydrogenase (G6PD); 3: 6-phosphogluconolactone (6PGL); 4: 6-phosphogluconate dehydrogenase (6PGDH); 5: ribulose-5-phosphate (Ru5P) isomerase (RPI); 6: Ru5P epimerase (RPE); 7: transketolase (TKT); 8: transaldolase (TALDO); 9: transketolase (TKT).

The first portion has three irreversible reactions where NADPH is generated in two of them (Table 6.1).[16] The first reaction begins with the dehydrogenation of G6P by G6P dehydrogenase (G6PDH) and leads to the formation of NADPH and 6-phosphogluconolactone. Subsequently, 6-phosphogluconolactone is hydrolyzed by phosphogluconolactonase (6PGL) into 6-phosphogluconate. A second NADPH is generated in the third reaction, where 6-phosphogluconolactone is oxidatively decarboxylated by 6-phosphogluconate dehydrogenase (6PGDH) into ribose-5-phosphate (Ru5P). The non-oxidative phase of the PPP comprises a

Table 6.1. Reactions in oxidative phase of the PPP.

Oxidative phase	Reactions	Enzyme
1	Glucose 6-phosphate + NADP$^+$ $\longrightarrow$ 6-Phosphoglucono-δ-lactone + NADPH + H$^+$	G6PD
2	6-Phosphoglucono-δ-lactone $\longrightarrow$ 6-Phosphogluconate	6PGL
3	6-Phosphogluconate + NADP$^+$ $\longrightarrow$ Ribulose-5-phosphate + NADPH + H$^+$ + CO$_2$	6PGDH

series of readily reversible reactions and thus results in the rearrangement of the carbon skeletons of numerous carbohydrates including xylulose-5-phosphate, sedoheptulose-7-phosphate, glyceraldehyde-3-phosphate, fructose-6-phosphate, and erythrose-4-phosphate (Figure 6.1).[9,17] Among those carbohydrates, additional glycolytic intermediates such as glyceraldehyde-3-phosphate and fructose-6-phosphate are recruited so that they can be converted into pentose phosphates again.

In the oxidative phase, G6PDH reaction is the rate-limiting step and is essentially irreversible. The most key regulation factor is the amount of NADP$^+$, the electron acceptor of G6PDH. The resulting NADPH during G6P dehydrogenation as well as NADP$^+$ will compete to bind with the active sites of G6PDH, leading to a decrease in enzyme activity. Hence, the ratio of NADPH/NADP$^+$ affects the activity of G6PDH directly.[18] On the other hand, in the non-oxidative phase of the PPP, all the reactions are reversible enabling cells to adapt themselves to the demands of carbohydrate metabolism and reducing equivalents with a great deal of flexibility. Two primary enzymes involved in the non-oxidative phase of the PPP are transaldolase and TKT, which are the link back to glycolysis.[19] TKT functions to transfer two-carbon groups from the substrates of the PPP and thus rearrange the carbon skeletons of numerous carbohydrates.[19] Moreover, TKT requires the coenzyme thiamine pyrophosphate (TPP) in the transfer reaction, while transaldolase does not. Thanks to the reversible reaction of TKT and transaldolase, the PPP and glycolysis pathway can be linked with each other and be regulated according to the demands of cells under different modes. There are three different modes in biological systems about the G6P flux[20]: (1) the demand of ribose-5-phosphate is higher than NADPH, which usually happens in rapidly dividing cells.[9] The PPP is diverted toward the generation of ribose-5-phosphate from both G6P in the oxidative phase and fructose-6-phosphate as well as glyceraldehyde-3-phosphate in the non-oxidative phase under the function of TKT and transaldolase. (2) The demand of ribose-5-phosphate and NADPH is balanced.[21] The oxidative phase of the PPP is dominant while ribose-5-phosphate and NADPH are generated. (3) The demand of NADPH is much higher than ribose-5-phosphate such as the process of fatty acid synthesis.[16] The oxidative phase of the PPP will be accelerated and the non-oxidative phase will be directed toward recruit fructose-6-phosphate and then converted back to G6P for the next oxidative phase to generate more NADPH. G6P will be finally oxidized to CO_2. Therefore, the reactions and products of the PPP can shift depending on the metabolic needs of a particular cell.[20]

There are two primary functions of the PPP. First of all, the PPP is the main pathway of cells to generate reducing equivalents in the form of NADPH, which serves as a cofactor and is essential for reductive biosynthesis reactions within cell.[22,23] The second function of the PPP is to provide the cell with carbohydrates of different structures such as ribose-5-phosphate and erythrose-4-phosphate.[24] Ribose is synthesized through either the oxidative phase or the non-oxidative phase of the PPP, which is an important component of RNA or DNA as well as numerous cellular intermediates such as ATP, ADP, AMP, CoA, and FAD. In addition, the PPP can also provide conditions for the conversion of diverse monosaccharides.

6.3. Metabolic Engineering Applications of the PPP for Chemical Synthesis in the Microbes

6.3.1. *Increased NADPH Concentrations for Enhanced Chemical Synthesis*

NADPH is an essential electron donor in almost all the organisms and plays an important role in biosynthesis, which attracts great interest in the industry. For example, the production of many natural products often involves NADPH-dependent enzymes.[25,26] However, many biosynthetic reactions and bioconversions are limited by low availability of NADPH, leading to the low yield of target products. Three pathways including the oxidative PPP (oxPPP), the isocitrate dehydrogenase step of the tricarbo-xylic acid (TCA) cycle, and the Entner–Doudoroff (ED) pathway are tra-ditionally considered as the major sources of NADPH.[27] However, with the development of modern science, many other ways of generating NADPH have been designed and characterized (Table 6.2).[28] Nevertheless, the PPP contributes immensely to NADPH pool and one of the most important roles of the PPP is to produce NADPH, which is generated in two of the steps in the oxidative phase.[8] As described above, several enzymes involved in the PPP are related to the yield of NADPH including glucose 6-phosphate dehydrogenase (G6PDH), 6-phosphogluconate dehydrogenase (6PGDH), and TKT.[16] Many studies have demonstrated that the overexpression of the enzymes involved in the PPP contributed to the high NADPH yield.[29] Hence, metabolic engineering of the PPP has been an effective way to improve the generation of NADPH and thus enhance the production of target products.

Polyhydroxybutyrate (PHB) is a form of polyester synthesized by bacteria acting as intracellular granules.[32,33] Three primary reactions are involved in the biosynthesis of PHB catalyzed by enzymes including β-ketothiolase, NADPH-dependent acetoacetyl-CoA reductase, and PHB synthase.[34] Therefore, NADPH plays an important role as a cofactor in the NADPH-dependent biosynthesis pathway. Lim *et al.* amplified the NADPH-related genes *zwf* and *gnd*, encoding G6PDH and 6PGDH in *Escherichia coli*, respectively.[35] Overexpression of the two genes increased the level of NADPH and thus enhanced the biosynthesis of PHB. The overexpression of TKT and transaldolase in the non-oxidative phase of the PPP, encoded by genes *tktA* and *talA* in *E. coli*, also increased NADPH yield as well as PHB accumulation.[36,37]

Besides PHB, the production of free fatty acid was also improved by the metabolic engineering of the PPP in *Aspergillus oryzae*.[38] NADPH is required in the *de novo* fatty acid synthesis as a cofactor. Tamano *et al.* overexpressed a predicted TKT gene of the PPP and increased produc-tivity.[38] It has been proved that the oxidative phase of the PPP was also the primary source of NADPH for lipid overproduction in *Y. lipolytica* as well as some oleaginous fungus such as *Mucor circinelloides*.[39,40]

Table 6.2. Overview of major NADPH-generating enzymes.[28]

	Enzyme	Pathway	Additional cofactors	Application[a]	$\Delta_r G'^{mb}$ (kJ/mol)
Enzymes coupled to central carbon metabolism	G6PDH	oxPPP, ED	n.a.	Y	-2.3 ± 2.6
	6PGDH	oxPPP	n.a.	Y	-6.0 ± 6.3
	IDH	TCA cycle	n.a.	Y	-10.7 ± 6.3
	ME	Anaplerotic nodes	n.a.	N	-3.1 ± 6.2
	GAPN	EMP	n.a.	Y	-36.1 ± 1.1
	NADP+-GAPDH	EMP	n.a.	Y	25.9 ± 1.0
	GDHs	Modified EDs	n.a.	Y	-2.4 ± 2.2
Enzymes not coupled to central carbon metabolism	STH	n.a.	NADH, FAD	Y	1.0 ± 0.7
	H+-TH	n.a.	NADH	Y	1.0 ± 0.7
	FNR	n.a.	FAD or FMN, Fd_{red}	N	-15.6 ± 11.7
	SH	n.a.	FAD or FMN	N	-16.5 ± 5.9
	NADK	n.a.	NTP or poly(P), NAD+ or NADH	Y	n.a.

[a]Application and usage of the enzyme in metabolic engineering to increase NADPH level and thus enhance bio-reaction processes.
[b]$\Delta_r G'^m$, Gibbs free energies calculated using the biochemical thermodynamics calculator eQuilibrator2.0.[28,30,31]

n.a., not applicable; Y, yes; N, no; G6PDH, glucose-6-phosphate dehydrogenase; 6PGDH, 6-phosphogluconate dehydrogenase; IDH, isocitrate dehydrogenase; ME, malic enzyme; GAPN, non-phosphorylating glyceraldehyde 3-phosphate dehydrogenase; NADP+-GAPDH, NADP+-dependent phosphorylating glyceraldehyde 3-phosphate dehydrogenase; GDH, glucose dehydrogenase; STH, energy-independent soluble transhydrogenase; H+-TH, energy-dependent or proton-translocating, membrane-bound transhydrogenase; FNR, ferredoxin: NADP+ oxidoreductase; SH, cytosolic NADP+-reducing hydrogenase; NADK, NAD+ kinase.

Lipids originating from microorganisms have attracted much interest since they are potential sources of biofuels and some of them contain high-value polyunsaturated fatty acids including docosahexaenoic acid (DHA), γ-linolenic acid, arachidonic acid (AA), and so on. Lipid biosynthesis is similar in diverse microorganisms as described above, which needs NADPH as a cofactor. NADPH is important in the process of lipid biosynthesis and is provided by several pathways including the PPP. It has been reported that G6PD is the major contributor of NADPH in *M. alpine*.[41] Song group found that *M. circinelloides* WJ11 could produce lipid up to 36% of CDW, while *M. circinelloides* CBS 277.49 could produce lipid less than 15% of CDW. The analysis proved that the activities of G6PD and 6PGD in WJ11 were higher than that in CBS 277.49.[42] Meanwhile, WJ11 had a higher carbon flux toward the PPP compared with CBS 277.49, indicating that PPP was vital for the generation of NADPH in the synthesis of lipids.[43] Hence, Zhao *et al.* overexpressed the genes encoding G6PD and 6PGD in *M. circinelloides*, leading to the increase of lipid content of CDW up to 30% with higher G6PD and 6PGD activities as well as mRNA levels compared with the control strain.[40]

NADPH is an important cofactor in the production of natural products such as β-carotene.[44,45] β-Carotene is synthesized through either mevalonate (MVA) pathway in eukaryotic cells or the 2-C-methyl-D-erythritol-4-phosphate (MEP) pathway in prokaryotic cells, both of which need NADPH as a cofactor. Zhao *et al.* introduced β-carotene synthetic genes into the chromosome of *E. coli* followed by the engineering of central metabolic modules for improving β-carotene production.[45] According to their study, β-carotene synthetic pathway in recombinant *E. coli* was divided into five modules, including β-carotene synthesis, MEP, ATP synthesis, TCA cycle, and pentose phosphate pathway. Engineering of each module resulted in the improvement of β-carotene production. β-Carotene yield increased 21%, 17%, and 39% after modulating the single gene of ATP synthesis, pentose phosphate pathway, and TCA modules, respectively. Moreover, the combination of TCA and PPP modulating led to a synergistic effect on increasing β-carotene yield up to 64%. Hence, the increasing NADPH supply through engineering the PPP was vital for the biosynthesis of natural products.

The metabolic engineering of the PPP was also helpful for the coproduction of hydrogen (H_2) and ethanol. Sekar and his colleagues increased the coproduction of H_2 and ethanol to the yields close to their theoretical maximum via activating the PPP in a *pfkA*-deficient *E. coli*.[46] Their aim was to solve the problem of low carbon-to-H_2 yield suffered in fermentative H_2 production by coproduction of H_2 and ethanol. The production processes of biofuels and biochemicals demand a very high NADPH, which can be supported by the PPP. Hence, the authors tried several approaches to enhance the PPP for improved coproduction of H_2 and ethanol. As we know, the production of H_2 and ethanol requires anaerobic conditions, under which most glucose is metabolized through the Embden–Meyerhof–Parnas (EMP) pathway. In this study, first of all, the EMP pathway was downregulated by *pfkA* deletion, which encodes the major phosphofructokinase isozymein *E. coli* so that the carbon flux could be diverted to the PPP. Subsequently, the PPP was activated by overexpression of the key enzymes in the PPP, including *zwf*, encoding G6PD and *gnd*, encoding 6PGD. The effect of differential expression of *zwf* and *gnd* on the coproduction of H_2 and ethanol was also studied because the conversion of the carbon flux to the PPP was not complete. The results indicated that the incomplete conversion of the carbon flux to the PPP did not result from the low activities of *zwf* and *gnd*, which meant that there were still many problems that need to be solved in the regulation of the PPP. Regardless of this, the authors successfully increased the yields of H_2 and ethanol via the metabolic engineering of the PPP together with the regulation of other metabolite pathways.

The availability of NADPH is also essential for various other industrially important products including amino acids, organic acids, proteins, antibiotics, and high-value metabolites. Engineering of the PPP, which is the major source of NADPH in organisms, plays a key role in the increased production of diverse chemicals (Figure 6.2).

6.3.2. *Enhanced Utilization of Xylose and Related Biomass via Metabolic Engineering of the PPP*

The utilization of renewable resource, which is helpful for sustainable development, has been drawing much attention for a long time. Lignocellulosic biomass, one of the most frequently used renewable

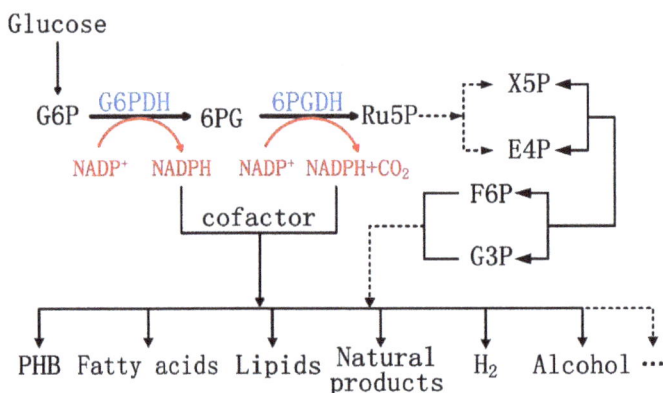

Figure 6.2.   Metabolic engineering of the PPP for the production of various chemicals. Cytosolic NADPH (red) was enhanced via overexpression of two key enzymes (blue) of the oxPPP. G6P, Glucose 6-phosphate; 6PG, 6-phosphogluconate; Ru5P, ribulose-5-phosphate; X5P, xylulose-5-phosphate; E4P, erythrose-4-phosphate; F6P, fructose-6-phosphate; G3P, glyceraldehyde-3-phosphate; PHB, polyhydroxybutyrate; G6PDH, glucose-6-phosphate dehydrogenase; 6PGDH, 6-phosphogluconate dehydrogenase; Dashed arrows meant multiple reactions.

resources, has been researched widely and considered as the most potential source for sustainable production since it is inexpensive, abundant, and available easily.[47] Lignocellulosic biomass, such as agricultural and forestry residues, contains mainly three kinds of biopolymers, namely cellulose, hemicellulose, and lignin.[47–49] Among the three kinds of biopolymers, hemicellulose, containing different polymers such as pentose, hexoses, and sugar acids, has been studied by many researchers since its decomposition is much easier than cellulose while the contents are relatively high.[49] As main components of hemicellulose, the efficient utilization of xylan as well as its hydrolysate xylose has remained a key problem over the past decades, especially in microbial productions when lignocellulosic hydrolysates are used as the carbon sources.[50]

Xylose metabolism in microbes is related to the PPP via the intermediate xylulose-5-phosphate, which can also be generated from xylose except for the traditional oxidative phase of the PPP. Xylose can be converted into xylulose via two pathways (Figure 6.3).[51–53] Most eukaryotes ferment xylose by a two-step redox reaction catalyzed by xylose reductase (XR)

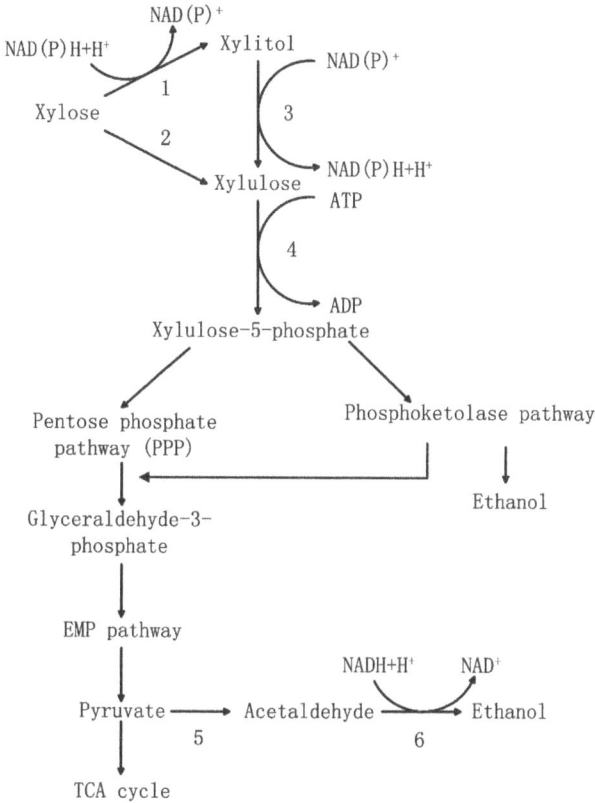

Figure 6.3. Xylose metabolism in microbes. 1: Xylose reductase (XR); 2: xylose isomerase (XI); 3: xylitol dehydrogenase (XDH); 4: xylulokinase (XK); 5: decarboxylase; 6: dehydrogenase.

and xylitol dehydrogenase (XDH) with xylitol as an intermediate.[54] In this conversion of xylose to xylulose, enzyme XR preferring NADPH while XDH exclusively using NAD+.[55,56] On the other hand, the conversion of xylose to xylulose can also be catalyzed by metalion-dependent enzyme xylose isomerase (XI) directly, which eliminates the excessive production of xylitol.[57] Subsequently, the resulting xylulose is phosphorylated to xylulose-5-phosphate (X5P) catalyzed by enzyme xylulokinase (XK), which is often overexpressed to drive more xylulose to X5P in many studies.[58,59] After that, X5P can be channeled through the PPP to glycolysis. In this way, xylose can be fermented as the carbon source of many microbes. Therefore, many studies have been focused on the study of xylose metabolism and enhanced utilization of xylose to produce chemicals such as ethanol through metabolic engineering of the PPP.

Lin *et al.* employed *Clostridium acetobytylcum*, an important industrial microorganism, which was capable of utilizing D-xylose and often used in the production of acetone, butanol, and ethanol to study the effect of the PPP on D-xylose utilization.[50] In their study, four genes in the genome of *C. acetobytylcum* including *tal*, *tkl*, *rpe*, and rpi were identified to be involved in PPP, encoding the four key enzymes transaldolase, TKT, ribose-5-phosphate isomerase, and ribose-5-phosphate epimerase, respectively, that catalyze PPP reactions related with xylose metabolism in microbes. The combined overexpression of these genes improved xylose utilization by *C. acetobytylcum* significantly with a solvent titer that is 42% higher than that of the wild-type strain. Moreover, enhanced xylose utilization resulting from the overexpression of the four genes in PPP increased the production of acetone, butanol, and ethanol compared with the original strain. This study offered a useful strategy to improve D-xylose utilization by *C. acetobytylcum* and the further applications on the production of various chemicals.

Saccharomyces cerevisiae, one of the most intensively studied eukaryotic model organism, has been considered as a fantastic host with immense potential for the production of biofuels, foods, natural products, and so on.[60,61] *S. cerevisiae* cannot utilize xylose directly like other yeast species.[56,62,63] Hence, *S. cerevisiae* has been extensively engineered to obtain the capability to use xylose efficiently.[53,64,65] For example, Zhou *et al.* obtained rapid xylose utilization in *S. cerevisiae* via metabolic engineering in xylose metabolism pathway and the PPP as well as evolutionary engineering.[52] In their study, a rational metabolic engineering of *S. cerevisiae* was conducted including the overexpression of two key genes in xylose utilization, namely xylose isomerase gene *XYLA* from *Piromyces* and xylulose kinase gene *XYL3* from *Pichia stipites*, and genes involved in the non-oxidative PPP to overcome potential limitations of xylose fermentation. The resulting engineered strain H131-A3 was subsequently used in a three-step process of evolutionary engineering. Sequential batch cultivation of H131-A3 in aerobic and anaerobic conditions was carried out followed by cultivation in a xylose-limited chemostat for evolutionary engineering to select strains with enhanced growth, xylose consumption rate, and ethanol production. After optimizing fermentation, the evolved strain H131-A3-AL^{CS} demonstrated significantly increased anaerobic growth rate and xylose consumption rate with higher ethanol yield compared with other evolved strains.

Regulation of xylose metabolism as well as the PPP is also important for their further study and applications. Dueber's group characterized xylose utilization in *S. cerevisiae*.[56] They simultaneously titrated heterologous expression of eight genes from *Scheffersomyces stipitis* involved in xylose utilization in *S. cerevisiae*, namely XR, XDH, XK, RPE, RKI, TKL, TAL, and PYK, to identify important factors in xylose utilization. Figure 6.4 showed the fungal catabolic pathway from xylose to pyruvate via the non-oxidative PPP (Figure 6.4).[56] A combinatorial approach based

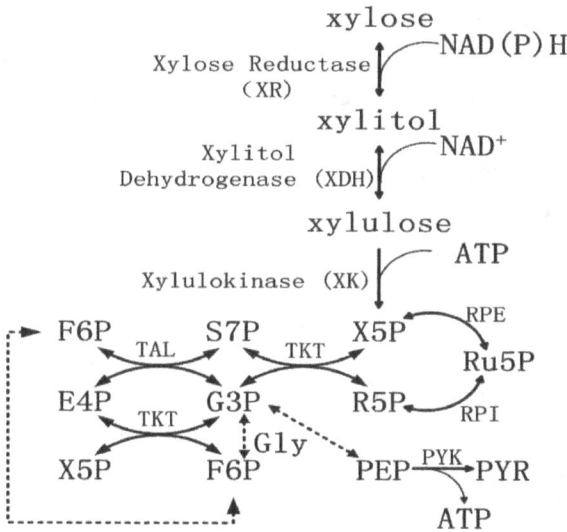

Figure 6.4. Catabolic pathway from xylose to pyruvate via the non-oxidative PPP in fungi.[56] X5P, Xylulose-5-phosphate; Ru5P, ribulose-5-phosphate; R5P, ribose-5-phosphate; S7P, sedoheptulose-7-phosphate; G3P, glyceraldehyde-3-phosphate; F6P, fructose-6-phosphate; E4P, erythrose-4-phosphate; PEP, phosphoenolpyruvate; PYR, pyruvate; Gly, glycolysis; RPE, Ru5P epimerase; RPI, Ru5P isomerase; TKT, transketolase; TAL, transaldolase; PYK, pyruvate kinase. Gly and PYK stand for the part of glycolysis.

on a set of five promoters was used to simultaneously regulate each gene and the promoter libraries were enriched under aerobic and anaerobic conditions or with a mutant XDH. After characterization, it was found that transaldolase activity in the PPP was limiting for aerobic xylose consumption and its expression primarily determined the xylose consumption together with XR expression. Therefore, the metabolic engineering of the PPP is quite essential for the utilization of xylose and the further chemical production in *S. cerevisiae*.

6.4. Future Potential Applications of Metabolic Engineering of the PPP

As an alternative pathway of glucose oxidation, the PPP has attracted the interest of many researchers for its multiple functions and connections with other pathways. The major functions of the PPP includes the yields of NADPH and production of ribose with some other applications such as oxidative stress prediction and cancer therapies.[9] The PPP is also connected to several pathways of carbohydrate metabolism like glycolysis, glucuronic acid pathway, and xylose metabolism. As discussed above, the PPP has been engineered for the characterization of its mechanism, the production of diverse chemicals, and the utilization of biomass. However, there are still some problems need to be solved during the process of utilizing the PPP for chemical production. First of all, the yield of NADPH through the PPP is not sufficient for some reactions in metabolism and thus affect the synthesis of target products. Even though many modifications have been tried to pull the carbon flux to the PPP for NADPH generation, it cannot meet the demand of many pathways until now. Second, the usage of biomass for sustainable production employing the PPP needs more research and will still take a long time for large-scale application since the studies are mainly focused on the combined metabolic engineering of the PPP and xylose metabolism, which is still limited for the utilization of abundant lignocellulosic biomass.

Hence, the PPP possesses more potential applications in the future when solving those problems. For instance, the combined metabolic engineering of the PPP and other pathways using synthetic biology and enzyme engineering tools. Li group studied the conversion of corncob hemicellulose hydrolysate or xylan to xylitol, one of the intermediates of xylose pathway, in *Candida tropicalis* employing different approaches.[66,67] Xylitol could be further consumed and enter the PPP. Aiming to improve the conversion of corncob hemicellulose hydrolysate to xylitol as well as the enzymatic corncob hemicellulose hydrolysate preparation efficiency, they developed an environment-friendly xylitol-producing bioprocess from hemicellulose hydrolysate with the optimization of xylanolytic enzyme production.[67] While for the conversion of xylan to xylitol, an integrated xylitol production pathway was generated via the construction of a new xylan fermentation pathway in engineered *C. tropicalis* BIT-Xol-1, which could convert xylose to xylitol, so that xylan could be used directly to produce xylitol without the addition of xylanolytic enzymes.[66] In this way, corncob hemicellulose hydrolysate can be used as a substrate directly for the production of xylitol, which can be connected with the PPP as described above. Therefore, the combined metabolic engineering of the PPP and xylitol production can be helpful for the production of various chemicals employing non-expensive and renewable biomass like hemicellulose as substrates as well as the further applications on large-scale production.

Even though a number of studies focused on the mechanism and regulations of the PPP, the large-scale applications of this pathway still need more efforts. The generation of NADPH and ribose via the PPP needs further studies for the sufficient supply as required. With the development of bioinformatics, synthetic biology, and biotechnology, more metabolic engineering of the PPP could be carried out for the production of high-value chemicals.

References

1. Sudesh, K., & Iwata, T. Sustainability of biobased and biodegradable plastics. *Clean-Soil Air Water* **36**, 433–442 (2008).
2. Sun, X. *et al.* Synthesis of chemicals by metabolic engineering of microbes. *Chem. Soc. Rev.* **44**, 3760–3785 (2015).
3. Jeandet, P., Vasserot, Y., Chastang, T., & Courot, E. Engineering microbial cells for the biosynthesis of natural compounds of pharmaceutical significance. *Biomed Res. Int.* **2013**, 780145 (2013).

4. Peralta-Yahya, P.P., Zhang, F., Del Cardayre, S.B., & Keasling, J.D. Microbial engineering for the production of advanced biofuels. *Nature* **488**, 320–328 (2012).

5. Nielsen, J., & Keasling, J.D. Synergies between synthetic biology and metabolic engineering. *Nat. Biotechnol.* **29**, 693–695 (2011).

6. Keasling, J.D. Manufacturing molecules through metabolic engineering. *Science* **330**, 1355–1358 (2010).

7. Romero-Rodriguez, A. *et al.* Transcriptomic analysis of a classical model of carbon catabolite regulation in *Streptomyces coelicolor*. *BMC Microbiol.* **16**, 77 (2016).

8. Wamelink, M.M.C., Struys, E.A., & Jakobs, C. The biochemistry, metabolism and inherited defects of the pentose phosphate pathway: A review. *J. Inherit. Metab. Dis.* **31**, 703–717 (2008).

9. Patra, K.C., & Hay, N. The pentose phosphate pathway and cancer. *Trends Biochem. Sci.* **39**, 347–354 (2014).

10. Horecker, B.L. The pentose phosphate pathway. *J. Biol. Chem.* **277**, 47965–47971 (2002).

11. Saggerson, D. Getting to grips with the pentose phosphate pathway in 1953. *Biochem. J.* **10**, 1042 (2009).

12. Dickens, F. Alternative routes of carbohydrate oxidation. *Br. Med. Bull.* **9**, 105–109 (1953).

13. Turner, J.F., & Turner, D.H. The regulation of glycolysis and the pentose phosphate pathway. *Biochem. Plants* **2**, 279–316 (2014).

14. Wood, T. Physiological functions of the pentose phosphate pathway. *Cell Biochem. Funct.* **4**, 241–247 (1986).

15. Eicks, M., Maurino, V., Knappe, S., Flügge, U.-I., & Fischer, K. The plastidic pentose phosphate translocator represents a link between the cytosolic and the plastidic pentose phosphate pathways in plants. *Plant Physiol.* **128**, 512–522 (2002).

16. Kruger, N.J., & von Schaewen, A. The oxidative pentose phosphate pathway: Structure and organisation. *Curr. Opin. Plant Biol.* **6**, 236–246 (2003).

17. Miosga, T., & Zimmermann, F.K. Cloning and characterization of the first two genes of the non-oxidative part of the *Saccharomyces cerevisiae* pentose-phosphate pathway. *Curr. Genet.* **30**, 404–409 (1996).

18. Nocon, J. *et al.* Increasing pentose phosphate pathway flux enhances recombinant protein production in *Pichia pastoris*. *Appl. Microbiol. Biotechnol.* **100**, 5955–5963 (2016).

19. Sprenger, G.A. Genetics of pentose-phosphate pathway enzymes of *Escherichia coli* K-12. *Arch. Microbiol.* **164**, 324–330 (1995).

20. Berg, J., Tymoczko, J., & Stryer, L. *Biochemistry*. WH Freeman, New York, USA (2002).
21. Stanton, R.C. Glucose-6-phosphate dehydrogenase, NADPH, and cell survival. *IUBMB Life* **64**, 362–369 (2012).
22. Pryke, J.A., & Rees, T. The pentose phosphate pathway as a source of NADPH for lignin synthesis. *Phytochemistry* **16**, 557–560 (1977).
23. Man, Z. *et al.* Systems pathway engineering of *Corynebacterium crenatum* for improved L-arginine production. *Sci. Rep.* **6**, 28629 (2016).
24. Juhnke, H., Krems, B., Kötter, P., & Entian, K.-D. Mutants that show increased sensitivity to hydrogen peroxide reveal an important role for the pentose phosphate pathway in protection of yeast against oxidative stress. *Mol. Gen. Genet.* **252**, 456–464 (1996).
25. Quartacci, M.F., Cosi, E., & Navari-Izzo, F. Lipids and NADPH-dependent superoxide production in plasma membrane vesicles from roots of wheat grown under copper deficiency or excess. *J. Exp. Bot.* **52**, 77–84 (2001).
26. Choi, Y.-N., & Park, J.M. Enhancing biomass and ethanol production by increasing NADPH production in *Synechocystis sp.* PCC 6803. *Bioresour. Technol.* **213**, 54–57 (2016).
27. Fuhrer, T., & Sauer, U. Different biochemical mechanisms ensure network-wide balancing of reducing equivalents in microbialmetabolism. *J. Bacteriol.* **191**, 2112–2121 (2009).
28. Spaans, S.K., Weusthuis, R.A., Van Der Oost, J., & Kengen, S.W. NADPH-generating systems in bacteria and archaea. *Front. Microbiol.* **6**, 742 (2015).
29. Poulsen, R.B. *et al.* Increased NADPH concentration obtained by metabolic engineering of the pentose phosphate pathway in *Aspergillus niger*. *FEBS J.* **272**, 1313–1325 (2005).
30. Flamholz, A., Noor, E., Bar-Even, A., & Milo, R. eQuilibrator — the biochemical thermodynamics calculator. *Nucleic Acids Res.* **40**, D770–D775 (2011).
31. Noor, E., Haraldsdóttir, H.S., Milo, R., & Fleming, R.M. Consistent estimation of Gibbs energy using component contributions. *PLoS Comput. Biol.* **9**, e1003098 (2013).
32. Wang, Y., Yin, J., & Chen, G.-Q. Polyhydroxyalkanoates, challenges and opportunities. *Curr. Opin. Biotechnol.* **30**, 59–65 (2014).
33. Wang, Y., Wu, H., Jiang, X., & Chen, G.-Q. Engineering *Escherichia coli* for enhanced production of poly (3-hydroxybutyrate-*co*-4-hydroxybutyrate) in larger cellular space. *Metab. Eng.* **25**, 183–193 (2014).
34. Chen, G.-Q., Hajnal, I., Wu, H., Lv, L., & Ye, J. Engineering biosynthesis mechanisms for diversifying polyhydroxyalkanoates. *Trends Biotechnol.* **33**, 565–574 (2015).

35. Lim, S.-J., Jung, Y.-M., Shin, H.-D., & Lee, Y.-H. Amplification of the NADPH-related genes zwf and gnd for the oddball biosynthesis of PHB in an *E. coli* transformant harboring a cloned phbCAB operon. *J. Biosci. Bioeng.* **93**, 543–549 (2002).

36. Song, B.-G., Kim, T.-K., Jung, Y.-M., & Lee, Y.-H. Modulation of *talA* gene in pentose phosphate pathway for overproduction of poly-β-hydroxybutyrate in transformant *Escherichia coli* harboring *phbCAB* operon. *J. Biosci. Bioeng.* **102**, 237–240 (2006).

37. Jung, Y.-M., Lee, J.-N., Shin, H.-D., & Lee, Y.-H. Role of *tktA* gene in pentose phosphate pathway on odd-ball biosynthesis of poly-β-hydroxybutyrate in transformant *Escherichia coli* harboring phbCAB operon. *J. Biosci. Bioeng.* **98**, 224–227 (2004).

38. Tamano, K., & Miura, A. Further increased production of free fatty acids by overexpressing a predicted transketolase gene of the pentose phosphate pathway in *Aspergillus oryzae FaaA* disruptant. *Biosci. Biotechnol. Biochem.* **80**, 1829–1835 (2016).

39. Wasylenko, T.M., Ahn, W.S., & Stephanopoulos, G. The oxidative pentose phosphate pathway is the primary source of NADPH for lipid overproduction from glucose in *Yarrowia lipolytica*. *Metab. Eng.* **30**, 27–39 (2015).

40. Zhao, L. *et al.* Role of pentose phosphate pathway in lipid accumulation of oleaginous fungus *Mucor circinelloides*. *RSC Adv.* **5**, 97658–97664 (2015).

41. Chen, H. *et al.* Identification of a critical determinant that enables efficient fatty acid synthesis in oleaginous fungi. *Sci. Rep.* **5**, 11247 (2015).

42. Tang, X. *et al.* Complete genome sequence of a high lipid-producing strain of *Mucor circinelloides* WJ11 and comparative genome analysis with a low lipid-producing strain CBS 277.49. *PLoS One* **10**, e0137543 (2015).

43. Zhao, L. *et al.* 13 C-metabolic flux analysis of lipid accumulation in the oleaginous fungus *Mucor circinelloides*. *Bioresour. Technol.* **197**, 23–29 (2015).

44. Chemler, J.A., Fowler, Z.L., McHugh, K.P., & Koffas, M.A. Improving NADPH availability for natural product biosynthesis in *Escherichia coli* by metabolic engineering. *Metab. Eng.* **12**, 96–104 (2010).

45. Zhao, J. *et al.* Engineering central metabolic modules of *Escherichia coli* for improving beta-carotene production. *Metab. Eng.* **17**, 42–50 (2013).

46. Sekar, B.S., Seol, E., Raj, S.M., & Park, S. Co-production of hydrogen and ethanol by *pfkA*-deficient *Escherichia coli* with activated pentose-phosphate pathway: Reduction of pyruvate accumulation. *Biotechnol. Biofuels* **9**, 95 (2016).

47. Zhou, C.-H., Xia, X., Lin, C.-X., Tong, D.-S., & Beltramini, J. Catalytic conversion of lignocellulosic biomass to fine chemicals and fuels. *Chem. Soc. Rev.* **40**, 5588–5617 (2011).
48. Brandt, A., Gräsvik, J., Hallett, J.P., & Welton, T. Deconstruction of lignocellulosic biomass with ionic liquids. *Green Chem.* **15**, 550–583 (2013).
49. Hendriks, A., & Zeeman, G. Pretreatments to enhance the digestibility of lignocellulosic biomass. *Bioresour. Technol.* **100**, 10–18 (2009).
50. Jin, L. *et al.* Combined overexpression of genes involved in pentose phosphate pathway enables enhanced D-xylose utilization by *Clostridium acetobutylicum*. *J. Biotechnol.* **173**, 7–9 (2014).
51. Chan, Y.K. *et al. Pentoses and Lignin.* Springer Berlin Heidelberg, Berlin, Germany, (1983).
52. Zhou, H., Cheng, J.S., Wang, B.L., Fink, G.R., & Stephanopoulos, G. Xylose isomerase overexpression along with engineering of the pentose phosphate pathway and evolutionary engineering enable rapid xylose utilization and ethanol production by *Saccharomyces cerevisiae*. *Metab. Eng.* **14**, 611–622 (2012).
53. Moysés, D.N., Reis, V.C.B., Almeida, J.R.M.d., Moraes, L.M.P.d., & Torres, F.A.G. Xylose Fermentation by *Saccharomyces cerevisiae*: Challenges and prospects. *Int. J. Mol. Sci.* **17**, 207 (2016).
54. Wohlbach, D.J. *et al.* Comparative genomics of xylose-fermenting fungi for enhanced biofuel production. *Proc. Natl. Acad. Sci. USA* **108**, 13212–13217 (2011).
55. Krahulec, S. *et al.* Fermentation of mixed glucose-xylose substrates by engineered strains of *Saccharomyces cerevisiae*: Role of the coenzyme specificity of xylose reductase, and effect of glucose on xylose utilization. *Microb. Cell Fact.* **9**, 16 (2010).
56. Latimer, L.N. *et al.* Employing a combinatorial expression approach to characterize xylose utilization in *Saccharomyces cerevisiae*. *Metab. Eng.* **25**, 20–29 (2014).
57. Liu, Y., Rainey, P.B., & Zhang, X.X. Molecular mechanisms of xylose utilization by *Pseudomonas fluorescens*: Overlapping genetic responses to xylose, xylulose, ribose and mannitol. *Mol. Microbiol.* **98**, 553–570 (2015).
58. Kim, S.R. *et al.* Rational and evolutionary engineering approaches uncover a small set of genetic changes efficient for rapid xylose fermentation in *Saccharomyces cerevisiae*. *PLoS One* **8**, e57048 (2013).
59. Karhumaa, K., Hahn-Hägerdal, B., & Gorwa-Grauslund, M.F. Investigation of limiting metabolic steps in the utilization of xylose by recombinant *Saccharomyces cerevisiae* using metabolic engineering. *Yeast* **22**, 359–368 (2005).

60. Legras, J.L., Merdinoglu, D., Cornuet, J., & Karst, F. Bread, beer and wine: *Saccharomyces cerevisiae* diversity reflects human history. *Mol. Ecol.* **16**, 2091–2102 (2007).

61. Stefanini, I. *et al.* Role of social wasps in *Saccharomyces cerevisiae* ecology and evolution. *Proc. Natl. Acad. Sci. USA* **109**, 13398–13403 (2012).

62. Hinman, N.D., Wright, J.D., Hogland, W., & Wyman, C.E. Xylose fermentation. *Appl. Biochem. Biotechnol.* **20**, 391–401 (1989).

63. Kötter, P., & Ciriacy, M. Xylose fermentation by *Saccharomyces cerevisiae*. *Appl. Biochem. Biotechnol.* **38**, 776–783 (1993).

64. Van Vleet, J., & Jeffries, T.W. Yeast metabolic engineering for hemicellulosic ethanol production. *Curr. Opin. Biotechnol.* **20**, 300–306 (2009).

65. Guo, W., Sheng, J., Zhao, H., & Feng, X. Metabolic engineering of *Saccharomyces cerevisiae* to produce 1-hexadecanol from xylose. *Microb. Cell Fact.* **15**, 24 (2016).

66. Guo, X. *et al.* A novel pathway construction in *Candida tropicalis* for direct xylitol conversion from corncob xylan. *Bioresour. Technol.* **128**, 547–552 (2013).

67. Li, Z., Guo, X., Feng, X., & Li, C. An environment friendly and efficient process for xylitol bioconversion from enzymatic corncob hydrolysate by adapted *Candidatropicalis*. *Chem. Eng. J.* **263**, 249–256 (2015).

Chapter 7

Mevalonate/2-Methylerythritol 4-Phosphate Pathways and Their Metabolic Engineering Applications

Xinxiao Sun[*,†] *and Qipeng Yuan*[*,†,‡]

* State Key Laboratory of Chemical Resource Engineering
Beijing University of Chemical Technology
Beijing 100029, China
†Beijing Advanced Innovation Center
for Soft Matter Science and Engineering
Beijing University of Chemical Technology
Beijing 100029, China
‡yuanqp@mail.buct.edu.cn

7.1. Introduction

Isoprenoids (also called terpenoids) are a diverse group of natural products comprising more than 35,000 identified compounds.[1] Depending on the number of isoprene units in the carbon skeleton, isoprenoids can be categorized into different groups (hemi-, mono-, sesqui-, di-, tri-, and tetra-terpenoids). Isoprenoids are functionally involved in many cellular processes such as photosynthesis, respiration, hormonal regulation of metabolism, regulation of growth and development, and defense against pathogen attack.[2] Isoprenoids also find wide applications as flavors (e.g.,

187

limonene and linalool), fragrances (e.g., citronellol and geraniol), nutraceuticals (e.g., astaxanthin and lycopene), pharmaceuticals (e.g., artemisinin, paclitaxel), colorants (e.g., β-carotene), and agrichemicals (e.g., gibberellins). Traditionally, the commercial supply of isoprenoids is largely dependent on the extraction from plant tissues. However, several drawbacks accompany this method such as low concentration, poor yields, high extraction and purification cost, damage on the ecological balance, and inconsistency of composition and concentration influenced by geographical location and weather.

Due to structural complexity, chemical synthesis of isoprenoids often suffers from side reactions and low yield although it proved to be feasible for some monoterpenes and carotenoids.[3] *In vitro* enzymatic production of isoprenoids using plant isoprenoid synthases is also impractical due to the high cost of enzymes and precursors, as well as poor conversion. The production of isoprenoids using metabolic engineered microorganisms is a promising method and has attracted increasing attention. A large amount of high-value isoprenoids may be produced from simple and cheap carbon sources. In addition, engineered microorganisms with tailored enzymes may produce non-natural isoprenoids with improved biological functions. Construction of productive cell factories relies on detailed information about the isoprenoid biosynthetic pathways and enzymes involved and also an understanding of the mechanisms of metabolic regulation.

Despite their structural diversity, isoprenoids are all derived from the same five-carbon precursors, isopentenyl diphosphate (IPP), and its isomer dimethylallyl diphosphate (DMAPP). There are two distinct pathways that can lead to IPP and DMAPP, which are the mevalonate (MVA) pathway and the 2-methylerythritol 4-phosphate (MEP) pathway. The MVA pathway was discovered in 1960s, and for a long time has been considered as the exclusive pathway to synthesize IPP and DMAPP. Until the 1990s, the existence of an alternative pathway, namely MEP pathway, was confirmed. While some bacteria and all eukaryotes in the cytoplasm synthesize IPP and DMAPP via the MVA pathway, most prokaryotes including *Escherichia coli* and plant plastids synthesize them through the MEP pathway.[4,5] In the following sections, we summarize the enzymatic and regulation information on the MVA/MEP pathways and recent progress on metabolic engineering for the production of functional isoprenoids.

7.2. The MVA Pathway

The MVA pathway starts with the condensation of two molecules of acetyl-CoA to generate acetoacetyl-CoA and the reaction is catalyzed by the acetoacetyl-CoA thiolase (ACT) (Figure 7.1). In the second step, acetoacetyl-CoA condenses with a second acetyl-CoA to generate 3-hydroxy-3-methylglutaryl-CoA (HMG-CoA), which is catalyzed by HMG-CoA synthase (HMGS). The crystal structures of HMGSs from different organisms, such as *Staphylococcus aureus*, *Enterococcus faecalis*, and *Brassica juncea* have been determined. Although bacterial and eukaryotic HMGSs showed little sequence similarity with each other, the amino acid residues related to the acetylation and condensation reactions are conserved.[6] In the third step, HMG-CoA reductase (HMGR) converts HMG-CoA into MVA, which is further phosphorylated at the C-5 position by mevalonate kinase (MK) to generate phosphomevalonate. In the following step, phosphomevalonate kinase (PMK) converts phosphomevalonate into diphosphomevalonate, which is further converted into IPP by mevalonate diphosphate decarboxylase (MDC). IPP is then converted into DMAPP by IPP: DMAPP isomerase (IDI). The isomerization leads to over a billion-fold increase in the electrophilicity of the isoprene unit and represents an essential activation step in isoprenoid biosynthesis.[7] There are two types of IDI reported. The type I enzyme is widely distributed in eukaryota and eubacteria, while the type II enzyme was discovered in the

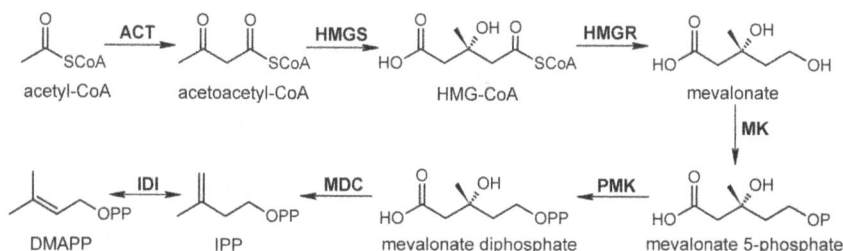

Figure 7.1. The metabolites and enzymes involved in the MVA pathway. ACT, acetoacetyl-CoA thiolase; DMAPP, dimethylallyl diphosphate; HMG-CoA, 3-hydroxy-3-methylglutaryl-CoA; HMGS, HMG-CoA synthase; HMGR, HMG-CoA reductases; IDI, IPP: DMAPP isomerase; IPP, isopentenyl diphosphate; MK, mevalonate kinase; PMK, phosphomevalonate kinase; MDC, diphosphomevalonate decarboxylase.

archaeon and streptomyces.[8,9] The two types of isomerases have distinct structures and different cofactor requirements, suggesting their different catalytic mechanisms.

The following steps are common for both MVA and MEP pathways. First, DMAPP is condensed with IPP to generate geranyl diphosphate (GPP), which is further combined with a second IPP to generate farnesyl diphosphate (FPP). The enzyme involved is called GPS/FPS. FPP is situated at the pivotal branch point, leading to sesqui-, di-, tri-, and tetra-terpenoids biosynthesis and is also a key regulation node of the MVA and MEP pathways. The MVA pathway is subjected to complex regulation, which is achieved mainly via multiple levels of feedback inhibition exhibiting effects on transcription, translation, enzyme activity, and protein stability.[10] Among the regulation points, HMGR and the distribution of FPP are two major ones.

7.3. The MEP Pathway

The MEP pathway is discovered relatively recently. The results of isotope labeling experiment suggested that in many eubacteria and plants, IPP and DMAPP are derived from pyruvate and glyceraldehyde 3-phosphate (G3P).[11,12] The full MEP pathway was elucidated through multidisciplinary approaches including biochemistry, bioinformatics, molecular biology, and organic chemistry. It comprises seven enzymatic steps (Figure 7.2). In the first step, pyruvate is condensed with G3P to generate 1-deoxy-D-xylulose 5-phosphate (DXP), which is catalyzed by DXP synthase (DXS). In the following step, DXP is converted into MEP by DXP reductoisomerase (DXR). MEP is then converted to 4-diphosphocytidyl-2-C-methyl-D-erythritol (CDP-ME) by MEP cytidylyltransferase (CMS) in the presence of CTP. Then, CDP-ME kinase (CMK) catalyzes the formation of 2-phospho-4-diphosphocytidyl-2-C-methyl-D-erythritol (CDP-MEP) from CDP-ME in the presence of ATP. CDP-MEP is further converted into 2-C-methyl-D-erythritol 2,4-cyclodiphosphate (ME-cDP) by ME-cDP synthase (MCS). ME-cDP is converted into hydroxymethylbutenyl diphosphate (HMBDP) catalyzed by HMBDP synthase (HDS). In the last step, HMDBP is reduced to IDP and DMADP by HMBDP reductase (HDR). The detailed information about the pathway enzymes is well described in Ref. [13].

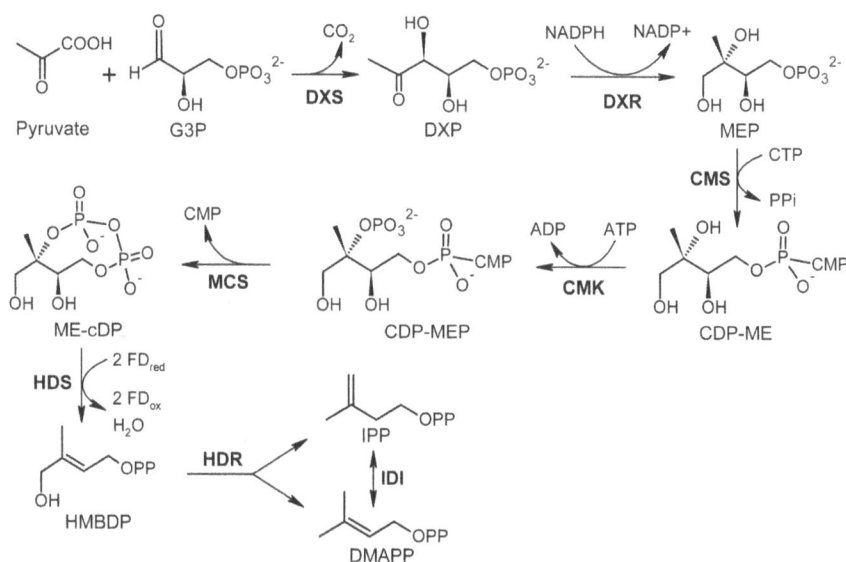

Figure 7.2. The metabolites and enzymes involved in the MEP pathway. G3P, glyceraldehyde 3-phosphate; DXP, 1-deoxy-D-xylulose 5-phosphate; MEP, 2-methylerythritol 4-phosphate; CDP-ME, 4-diphosphocytidyl-2-C-methyl-D-erythritol; CDP-MEP, 2-phospho-4-diphosphocytidyl-2-C-methyl-D-erythritol; ME-cDP, 2-C-methyl-D-erythritol 2,4-cyclodiphosphate; HMBDP, hydroxymethylbutenyl diphosphate; DXS, DXP synthase; DXR, DXP reductoisomerase; CMS, MEP cytidylyltransferase; CMK, CDP-ME kinase; MCS, ME-cDP synthase; HDS, HMBDP synthase; HDR, HMBDP reductase.

As mentioned above, the MVA/MEP pathways can lead to a variety of functional isoprenoids. In the following section, we summarize recent progress on the biosynthesis of some representative isoprenoids using metabolic engineered microorganisms (Figure 7.3).

7.4. Metabolic Engineering Applications of MVA/MEP Pathways

7.4.1. *Metabolic Engineering for the Production of Hemiterpenoid Isoprene*

Isoprene (2-methyl-1,3-butadiene) is the simplest but the most important member of the isoprenoid family. It is mainly used to produce artificial

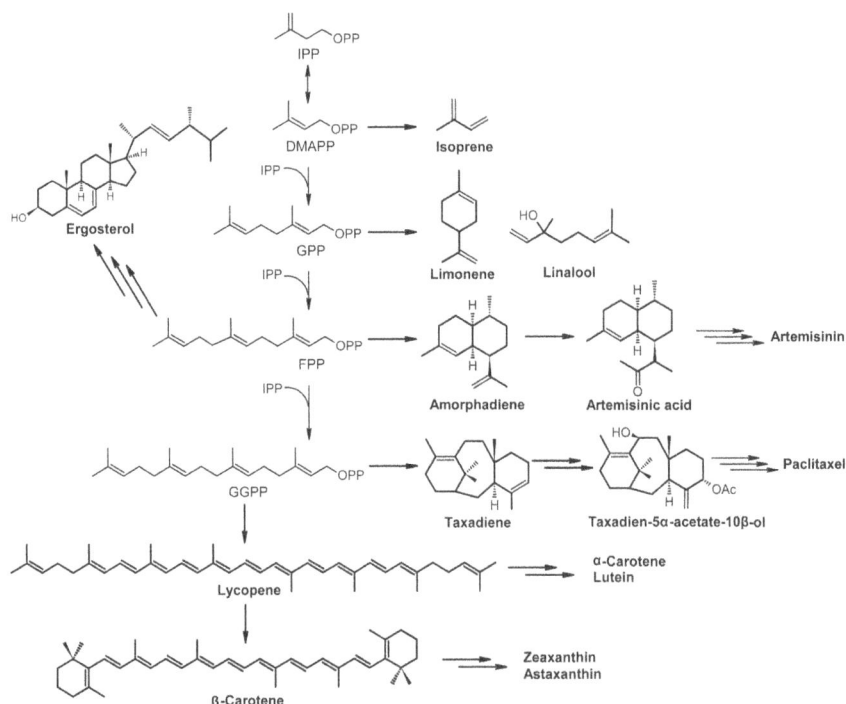

Figure 7.3. Biosynthetic pathways of different classes of isoprenoids. GPP, geranyl diphosphate; FPP, farnesyl diphosphate; GGPP, geranylgeranyl diphosphate.

rubber, which has an annual demand of over 20 million tons.[14] Currently, isoprene is manufactured primarily by cracking petroleum, which is unsustainable and frequently influenced by oil price. Many organisms including plants, bacteria, and humans can produce isoprene naturally. However, most of the isoprene is directly emitted into the atmosphere, which is impractical for recovering.[15] Although IPP and DMAPP can be polymerized by many plants into latex that is used for natural rubber production, it can meet only a small fraction of the demand in the world market. Therefore, it is of great importance to establish a sustainable and cost-effective process for isoprene production. As a model host strain, *E. coli* was extensively engineered to produce isoprene.[16] *E. coli* uses MEP pathway to synthesize IPP and DMAPP. The pathway is under tight metabolic regulation and the flux through this pathway is limited. Compared with the native MEP pathway, a foreign MVA pathway is free from the tight control and promisingly can lead to efficient production of isoprene. Therefore, a hybrid MVA pathway was constituted by expressing seven genes from several microorganisms. The *Enterococcus faecalis mvaE* and *mvaS* genes were cloned on a plasmid and are responsible for converting acetyl-CoA into mevalonate. *Saccharomyces cerevisiae* genes encoding MK, PMK, MDC, and IDI were organized as one operon and integrated into the chromosome, responsible for converting MVA to DMAPP.[16] The activity of MK is feedback inhibited by FPP and represents a bottleneck for isoprene production. To solve this problem, a feedback-resistant MK from *Methanosarcina mazei* was expressed, which presumably is a key step toward high-yield isoprene production. After constructing an efficient MVA pathway, the next step was to choose an isoprene synthase (IspS) with good kinetic properties. Only three-plant IspS had been characterized and none of them met the standards. *Populus alba* IspS was subject to a rational mutation by protein engineering. With an engineered version of IspS, the titer and yield of isoprene were significantly improved. In addition, HMGR requires NADPH as the cofactor. To increase its supply, the pentose phosphate pathway, which is the main source of NADPH, was enhanced by overexpressing *E. coli* phosphogluconolactonase gene (*ybhE*). With the extensively engineered strain, a large-scale continuous fermentation process was developed and a high percent of isoprene was recovered from the off-gas. The titer, volumetric productivity, and yield reached over 60 g/L, 2 g/L/h, and 11% (g/g glucose), respectively.[16]

7.4.2. *Metabolic Engineering for the Production of Monoterpenes*

Monoterpenes are important components of the essential oils of herbal plants. A variety of monoterpenes like menthol, D-limonene, D-carvone, and citral contribute to various odors and are widely used in flavor and fragrance industries.[17] Due to their relatively simpler structures, functional monoterpenes are mainly manufactured via chemical synthesis. Researchers also pursued the possibility to produce monoterpenes using metabolic engineered microorganisms. Monoterpenes are biosynthesized from GPP, and the enzyme catalyzing the committed step is called monoterpene synthase (MTS). Grafting MTSs into an efficient GPP biosynthetic pathway can lead to accumulation and secretion of monoterpenes. Various MTSs have been identified in plants. *Abies grandis* GPP synthase, and *Mentha spicata* L-limonene synthase genes were expressed in *E. coli* and nearly 5 mg/L of limonene was produced.[18] Constitutively expressing limonene synthase gene resulted in more than 450% increase of limonene contents in spike lavender compared with non-transformed controls.[19] Introducing an (S)-linalool synthase gene into a recombinant *S. cerevisiae* led to linalool excretion.[20] The titer of linalool was doubled when MVA pathway was enhanced by expressing a truncated HMGR (tHMGR).[21]

7.4.3. *Metabolic Engineering for the Production of Sesquiterpenes*

Sesquiterpenes are derived from the common precursor FPP, and sesquiterpene synthases catalyze the committed steps. Sesquiterpenes exhibit various physiological functions in plants and some of them have important pharmaceutical applications, among which artemisinin is the most well known. Here, we will take artemisinin as an example to show the promising potential of metabolic engineering for sustainable production of important sesquiterpenes.

Artemisinin, a sesquiterpene molecule containing 15 carbon atoms and a lactone endoperoxide, is an effective antimalarial drug. Malaria poses a great threat to human health, especially to those in the developing countries. In 2015, an estimated 214 million new cases of malaria were reported and 438,000 associated deaths occurred.[22] Millions of people are still not acquiring the services to prevent and treat malaria. Malaria is caused by *Plasmodium* parasites, which have developed widespread resistance to the formerly used antimalarial drugs (such as chloroquine and sulfadoxine–pyrimethamine). In 1970s, a research program was started by the Chinese government, aiming to find new antimalarial drugs. The *Artemisia annua* extracts were found to be effective against chloroquine-resistant *P. falciparum*. The active compound was later identified as artemisinin, and its structure was subsequently determined. Recommended by the WHO, now artemisinin derivatives are primary components of antimalarial combination therapies (ACTs). ACTs are more effective at shortening treatment times and reducing the potential for the emergence of artemisinin-resistant parasites. The supply of artemisinin relied exclusively on extraction from the plant *A. annua* and its price and availability are affected by weather and cultivation acreage, which fluctuate dramatically with time. To stabilize the supply and decrease the cost of artemisinin, the semisynthetic artemisinin project was proposed. The overall aim of this project was to engineer a microorganism to produce an artemisinin precursor at high titers, rates, and yields, followed by chemical conversion to artemisinin. The success of this project depends on a complete understanding of its biosynthesis in *A. annua*, the availability of rapid tools of microbial genetic manipulation, and the application of pharmaceutical chemistry.

In *A. annua*, artemisinin is synthesized via the MVA pathway. The common precursor FPP is converted into amorphadiene by amorphadiene synthase (ADS).[23] The oxidation of amorphadiene to artemisinic acid is catalyzed by a cytochrome P450 enzyme (CYP450) CYP71AV1.[24] CYP71AV1 is located in the trichomes of *A. annua* and its activity requires a cognate reductase CPR1. As the following steps were not characterized at that time, artemisinic acid was chosen as the suitable precursor for further chemical conversion. At first, *E. coli* was used as the host for artemisinic acid biosynthesis. The native MEP pathway was enhanced by overexpressing the rate-limiting enzymes to increase the precursor supply. However, this strategy was unsuccessful and the titer of amorphadiene is only in low milligrams per liter level. The limited titer was presumed to be due to two main reasons. First, the native MEP pathway is under tight regulation and is difficult to achieve high flux. Second, plant enzymes were poorly expressed in *E. coli*. Thus, a foreign MVA pathway was introduced into *E. coli*.[25] The pathway was divided into the top pathway and the bottom pathway, which comprises three genes (*atoB*, *ERG13*, and *tHMG1*) and five genes (*idi*, *ispA*, *MVD1*, *ERG8*, and *ERG12*), respectively. In addition, plant ADS was codon-optimized for better expression.[25] Recombinant *E. coli*, with the heterologous MVA pathway and codon-optimized ADS, produced 0.5 g/L of amorphadiene in a two-phase partitioning bioreactor.[26] To further identify the limiting steps, exogenous MVA was supplemented into the cell cultures of *E. coli* strain harboring the full amorphadiene biosynthetic pathway. The result showed that the addition of MAV significantly improved amorphadiene production, indicating that the top pathway is rate limiting. Therefore, to increase the expression of the three genes in the top pathway, the lac promoter was replaced with a stronger araBAD promoter. However, increased expression of the top pathway genes led to poor cell growth. Systematic analysis indicated that the expression of the three enzymes was imbalanced, leading to the accumulation of HMG-CoA, which inhibited cell growth.[27] To prevent HMG-CoA accumulation, the top pathway was balanced using libraries of tunable intergenic regions (TIGRs). With this approach, a strain with a 7-fold increase in MVA production was obtained.[28] It had been shown that high-titer production of MVA could be achieved by expressing a type II HMGR from *Enterococcus faecalis* in *E. coli*. Accordingly, several type II HMGRs were tested to replace yeast HMGR and the one from *Staphylococcus aureus* led to the highest production of amorphadiene. Replacing yeast HMGS with its counterpart from *S. aureus* further increased amorphadiene production. After obtaining the engineered strain with excellent performance, a novel fed-batch fermentation process with nitrogen and carbon limitation was developed, resulting in over 25 g/L of amorphadiene.[29] However, difficulties were encountered in the oxidation of amorphadiene to artemisinic acid in engineered *E. coli*.

Although the production of high titer of amorphadiene in *E. coli* was a pivotal step forward, artemisinic acid is the preferred precursor for the following chemical conversion. However, the oxidative reaction from amorphadiene to artemisinic acid is catalyzed by a eukaryotic CYP450 CYP71AV1, which is typically unsuitable to be expressed in *E. coli*. Recombinant *E. coli* expressing CYP71AV1 could only produce about 1 g/L of artemisinic acid which is too low for industrial application.[30]

Compared with *E. coli*, *S. cerevisiae* is more suitable for the functional expression of CYP71AV1. At first, *S. cerevisiae* S288C was used as the host strain as its genome had been sequenced. Yeast overexpressing ADS alone produced 4.4 mg/L of amorphadiene. To increase precursor supply, tHMGR was overexpressed, and amorphadiene production was improved approximately 5-fold. Squalene synthase (ERG9) catalyzes the key step of sterol biosynthesis from FPP. To downregulate the expression of ERG9, its native promoter was replaced with a methionine-repressible promoter (P_{MET3}). This modification led to an additional 2-fold increase in amorphadiene production. Expressing *upc2-1*, a global transcription factor regulating sterol biosynthesis further improved the titer to 105 mg/L. Overexpressing an additional tHMGR and FPP synthase (FPS) resulted in a strain able to produce 153 mg/L of amorphadiene, which was still much lower than that obtained in *E. coli* (25 g/L).[31] Further expressing CYP71AV1 along with its cognate reductase CPR1 led to over 100 mg/L of artemisinic acid. Incorporation of ADS, CYP71AV1, and CPR1 on a single plasmid led to production of 2.5 g/L of artemisinic acid using a galactose-based fermentation process.[32] To further improve the titer, an alternative *S. cerevisiae* strain CEN.PK2 was used to replace the previously used strain S288C. Compared with S288C, CEN.PK2 sporulates more sufficiently and has characteristics desirable for industrial fermentation. With the same genetic modifications, the two strains showed comparable artemisinic acid production in small-scale cultures. Strain CEN.PK2 was further modified by overexpressing all MVA pathway genes as far as *ERG20* encoding FPS. In addition, glucose replaced galactose as the carbon source to reduce the cost. Compared with the S288C-derived strain, the engineered CEN.PK2 strain produced 5-fold higher concentrations of amorphadiene. Under optimized fermentation conditions, the titer of amorphadiene reached 40 g/L.[33] However, introducing *CYP71AV1* into the amorphadiene-overproducing strain failed to obtain a satisfactory artemisinic acid production. Further investigation showed that artemisinic acid production led to a severe decrease in cell viability, which is caused by increased oxidative stress. It was reported that incomplete coupling between CYP450s and their reductases can lead to the release of reactive oxygen species. In liver microsomes, the abundance of the CYP450 is generally higher than its reductase. However, in the yeast strain, CYP71AV1 and its reductase CPR1 were expressed at similarly high concentrations. Therefore, the expression of CPR1 was reduced, resulting in improved viability but decreased production of artemisinic acid. It was reported that cytochrome *b*5 can interact with some CYP450s and increase their reaction rate. Thus, a cytochrome *b*5 from *A. annua* was introduced in the CEN.PK2 strain, resulting in increased production of artemisinic acid. However, this strain also produced a high amount of artemisinic aldehyde, which is a highly active and presumably toxic intermediate.

To solve this problem, the *A. annua* artemisinic aldehyde dehydrogenase (ALDH1) and an NAD-dependent artemisinic alcohol dehydrogenase (ADH1) were expressed, resulting in a significant increase in artemisinic acid production. Finally, the *GAL80* gene was deleted to make the expression of pathway enzymes independent of galactose. With an optimized fermentation process, the final strain was able to produce 25 g/L of artemisinic acid.[34] A chemical process was also developed to convert the purified artemisinic acid into artemisinin.

7.4.4. *Metabolic Engineering for the Production of Diterpenes and Triterpenes*

Diterpenes and triterpenes also comprise numerous important compounds, such as paclitaxel and glycyrrhizin. Paclitaxel is a tetracyclic diterpene and has been used as an effective drug against breast, lung, and non-small cell cancers. Its annual market value is over $1 billion.[35] Paclitaxel was originally discovered from the bark of *Taxus brevifolia*, which is still the main source of commercial paclitaxel. However, the content of paclitaxel in the bark is extremely low, and the extraction and purification processes are laborious and suffer from low yield. For example, to obtain sufficient dosage for one patient, two to four grown trees need to be sacrificed. Therefore, researchers investigated alternative ways to produce paclitaxel, which include chemical synthesis, semisynthesis, plant cell culture, and recombinant microbial systems. Due to its structural complexity, total chemical synthesis of paclitaxel requires 35–51 steps, and the yields are less than 0.4%.[36] Later on, baccatin III, the biosynthetic intermediate of paclitaxel, was isolated from other plant sources and chemically converted to paclitaxel. Although this approach and the plant cell culture approach have decreased the immoderate exploitation of the yew tree, these plant-based processes are still limited by productivity and scalability. In recent years, with the partial elucidation of the biosynthetic pathway, researchers attempted to produce paclitaxel precursor using engineered microorganisms.

The biosynthesis of paclitaxel starts from the common diterpenoid precursor geranylgeranyl diphosphate (GGPP) and the pathway contains at least 19 steps.[37] The committed step was the formation of taxadiene from GGPP, which is catalyzed by taxadiene synthase (TS). Taxadiene is then hydroxylated at the 5-position by the CYP450 taxadiene-5α-hydroxy-lase (T5αH) to generate taxadiene-5α-ol. Taxadiene-5α-ol was further converted to paclitaxel in multiple steps, and most of the enzymes involved are still not characterized. Compared with artemisinin, microbial production of paclitaxel is more challenging. In 2001, *E. coli* strain was engineered to produce taxadiene and the native MEP pathway was used to supply IPP. In plants, TS contains a signal peptide in its N-terminal to facilitate transportation into vacuoles. Because *E. coli* lacks interior membrane structures and the MEP pathway is located in the cytosol, the signal peptide was removed, resulting in a truncated TS (tTS). Genes encoding IDI, GGPP synthase (GGPS), and tTS were overexpressed and 0.5 mg/L of taxadiene was produced. The MEP pathway was enhanced by overexpressing DXS and the titer was improved to 1.3 mg/L.[38] To further increase titers of taxadiene, a multivariate-modular approach was developed to balance the metabolic pathway. The taxadiene biosynthetic pathway was partitioned into two modules, which are separated at IPP. The upstream module comprises four rate-limiting genes in the MEP pathway, while the downstream module comprises two heterologous genes encoding GGPS and tTS. The expressing levels of the upstream and downstream pathways were balanced by varying the promoter strength and gene copy numbers. This strategy proved to be very effective and yielded 1.02 g/L taxadiene in fed-batch bioreactor fermentations.[36]

The pathway was further extended to produce taxadiene-5α-ol by introducing an engineered T5αH. To achieve its functional expression in *E. coli*, T5αH was codon-optimized and its N terminal was modified by transmembrane engineering while its C-terminal was fused with *Taxus* CYP450 reductase (TCPR). One of the chimera enzymes generated was able to efficiently carry out the oxidation step. With this enzyme, the optimized strain produced up to 58 mg/L taxadiene-5α-ol.[36]

S. cerevisiae was also used as a host for taxadiene biosynthesis. When an N-terminal truncated TS from *Taxus canadensis* (TcTS) was expressed, there was no detectable taxadiene produced. The native supply of GGPP was presumed to be insufficient to support taxadiene biosynthesis. Therefore, *T. canadensis* GGPS (TcGGPS) was introduced, resulting in the production of 204 μg/L of taxadiene. To increase supply of the building blocks, the yeast native MVA pathway was enhanced by overexpressing tHMGR, leading to a 50% increase in taxadiene accumulation. Inhibiting sterol production by expressing upc2-1 was not beneficial to taxadiene biosynthesis. Western blot results showed that TcGGPS was poorly expressed in yeast, probably due to the unfavorable codon usage. To solve this problem, TcGGPS was replaced by its counterpart from *S. acidocaldarius*. This modification led to more than a 100-fold increase in GGPP level (27.6 mg/L), but only marginal increases in taxadiene production. Thus, TS was supposed to be rate-limiting in taxadiene production. Northern and Western blots showed that TcTS was also poorly expressed, due to the existence of several rare arginine codons. Replacement of the wild-type TS with the codon-optimized version resulted in about 40-fold increase in taxadiene levels (8.7 mg/L).[39]

As indicated above, taxadiene production is more efficient in *E. coli* than in *S. cerevisiae*. However, for functional expression of plant CYP450s, *S. cerevisiae* is a better host as it has endoplasmic reticulum that is compatible with the membrane signal domains of CYP450s. Thus, an *E. coli–S. cerevisiae* coculture system was developed, in which taxadiene produced by engineered *E. coli* diffused into *S. cerevisiae* and was subsequently converted to a monoacetylated, deoxygenated taxane. To coordinate the growth of the two microorganisms, a mutualistic interaction was designed. For this purpose, xylose instead of glucose was used as the carbon source. *E. coli* can metabolite xylose and excrete acetate. On the other hand, *S. cerevisiae* cannot utilize xylose but can metabolite acetate, which removes the inhibitory effect of acetate on the growth of *E. coli*. By optimizing cultivation conditions and gene expression, the titer of the target compound reached 1 mg/L.[40]

Microbial production of triterpenes has not been active, since enzymes involved in their biosynthesis are not well characterized. *E. coli* was engineered to produce squalene by introducing heterologous squalene/phytoene synthases and FPS, and overexpressing its native DXS and IDI genes. The resultant strain produced 11.8 mg/L of squalene.[41] In another study, a gene encoding a CYP450 monooxygenase was identified from *Bupleurum falcatum* and designated CYP716Y1. Phylogenetic analysis showed that CYP716Y1 belongs to an enzyme family involved in triterpene saponin biosynthesis. *In vivo* assay further confirmed that it catalyzes the C-16α hydroxylation of β-amyrin. Combined expression of CYP716Y1 with oxidosqualene cyclase, CYP450, and glycosyltransferase genes enabled accumulation of mono-glycosylated saponins in *S. cerevisiae*.[42]

7.4.5. *Metabolic Engineering for the Production of Carotenoids*

Carotenoids (e.g., lycopene, β-carotene, astaxanthin, and lutein) are an important class of tetraterpenoid compounds naturally produced by plants and microorganisms. Most carotenoids present natural colors, from yellow to red. More importantly, they exhibit promising antioxidative activity, which is beneficial to human health. Because of these properties, carotenoids have been widely used as colorants, food additives, nutraceuticals, and pharmaceuticals. In 2010, the global market of carotenoids was estimated to be \$1.2 billion, and it is expected to reach \$1.4 billion in 2018.[43] Currently, the commercial carotenoids are primarily manufactured by chemical synthesis and extraction from vegetables. These approaches have inherent disadvantages. Chemical synthesis is environmentally unfriendly and cannot fulfill people's increasing demand for natural products. And extraction from plants is frequently influenced by the seasonal and geographic variability. Therefore, since the 1990s, researchers have been trying to harness microorganisms to produce carotenoids. Carotenogenic microbes such as *Blakeslea trispora* and *Xanthophyllomyces dendrorhous* can naturally accumulate carotenoids. Due to the lack of genetic manipulation tools, the performance of these microbes was mainly improved by traditional breeding strategies like mutagenesis and careful optimization of the cultivation conditions.

The elucidation of the carotenoid biosynthetic pathways and genes involved enables people to engineer non-carotenogenic model organisms (mostly *E. coil* and *S. cerevisiae*) for carotenoid production. Like other isoprenoids, carotenoids are synthesized via the MEP or MVA pathway. Two molecules of GGPP are condensed by phytoene synthase (CrtB) to form colorless C40 phytoene, which is further desaturated by phytoene desaturase (CrtI) to produce red-colored lycopene. Cyclic carotenoids are synthesized by the cyclization of one or both end groups of lycopene. For example, a lycopene β-cyclases (CrtY) cyclizes both ends of lycopene to make β-carotene, which in turn is converted into canthaxanthin and zeaxanthin by ketolase (CrtW) and β-hydroxylase (CrtZ), respectively, and finally to astaxanthin by their cooperative actions.

E. coli is suitable for functional expression of most of the carotenogenic genes. Extensive studies have been conducted to engineer *E. coli* for carotenoid production. The strategies used to improve carotenoid production include enhancing native MEP pathway/introducing foreign MVA pathway, balancing the supply of precursors, enhancing prenyl diphosphate pathway, and increasing ATP and NADPH supplies.[44] Individual overexpression of DXS, DXR, and IDI effectively increased carotenoid production, indicating that the steps catalyzed by these enzymes are rate-limiting for isoprenoid biosynthesis. Coexpression of these enzymes has a synergetic effect and increases the yields more significantly.[45] The MEP pathway starts by the condensation of pyruvate and G3P, two metabolites of glycolysis pathway. IPP synthesis requires pyruvate and G3P in equal amounts. However, in the cytosol pyruvate is more abundant than G3P. Redirecting flux from pyruvate back to G3P increased lycopene production.[46] Expression of pathway genes on multicopy plasmids often causes metabolic burden and genetic instability of host strains. To solve this problem, the native promoter of the chromosomal MEP pathway genes was replaced with the stronger T5 promoter. The resultant *E. coli* strain produced 6 mg/g DCW of β-carotene.[47] In 2013, Zhao *et al.* reported the systematic optimization of β-carotene production in *E. coli*.[49] The full pathway was divided into five modules, including β-carotene synthesis, MEP, ATP synthesis, TCA cycle, and pentose phosphate pathway (PPP). A starting strain was constructed by integrating the *crtEXYIB* gene operon into the chromosome. Gene expression was modulated by multiple regulatory parts with different strengths. Engineering MEP module led to 3.5-fold increase of β-carotene yield, while engineering β-carotene synthesis module led to another 3.4-fold increase. To produce one β-carotene, 8 ATP, 8 CTP, and 16 NADPH are consumed. To increase the supply of these cofactors, ATP synthesis, TCA cycle, and PPP were modulated and β-carotene yield increased to 21%, 17%, and 39%, respectively. Combined engineering of TCA and PPP modules exhibited a synergistic effect on β-carotene production, resulting in 64% increase of β-carotene yield over a high-producing starting strain. Fed-batch fermentation of the best strain produced 2.1 g/L β-carotene with a yield of 60 mg/g DCW.[48]

As all the sesquiterpenes, diterpenes, triterpenes, and tetraterpenes compete for the common precursor FPP, it is crucial to redirect FPP flux toward the target pathway to achieve efficient production of one desired type of isoprenoid. The previous strategies used include overexpressing the genes of the target pathway and downregulating the genes of the competing pathways. For carotenoid production in *S. cerevisiae*, the sterol pathway is a major competing pathway. However, as ergosterol is a necessary component of cell membrane, blocking this pathway is lethal to the cell. To divert the flux from sterol synthesis, the activity of the key enzyme ERG9 was inhibited with chemical inhibitors or the promoter of its encoding gene was replaced with a weak repressible promoter. However, early overexpression of pathway enzymes and early inhibition of ERG9 may exert metabolic stress and impair cell growth. To circumvent these drawbacks, a sequential control strategy was adopted to construct engineered yeast for lycopene production.[50] The centerpiece of the strategy is controlling gene expression by two promoters that are response oppositely to glucose concentration. The squalene pathway gene was controlled by glucose-inducible promoter P_{HXT1} while the carotenoid pathway genes were controlled by glucose-repressible promoter P_{GAL}. In the early cultivation stage with high concentration of glucose, the carotenoid pathway was repressed while the squalene pathway was activated, allowing for normal cell growth. Along with the consumption of glucose, the situation was gradually reversed. After 120 h of high-density fermentation, the engineered strain produced 1,156 mg/L of carotenoid with a yield of 20.79 mg/g DCW.[49]

Bioinformatics analysis indicates that a silent lycopene biosynthetic gene cluster exists in *Streptomyces avermitilis*. Bai *et al.* characterized a panel of native or synthetic promoters and RBSs using a flow cytometry-based quantitative method. Seven synthetic promoters with gradient strength were inserted upstream of the silent cluster. A correlation between lycopene accumulation and promoter strength was observed and the highest yield reached 82 mg/g DCW.[50]

7.5. Conclusion and Future Perspectives

Isoprenoids, a vast group of natural metabolites having various potential applications, are derived from the common MVA/MEP pathways. In the past decades, significant progress has been made in engineering

MVA/MEP pathways for the biosynthesis of functional isoprenoids. Now, artemisinic acid can be produced efficiently using engineered yeast and used for semisynthesis of artemisinin. The biosynthetic production of isoprene and farnesene is also close to commercial viability. However, the biosynthetic pathways of many other isoprenoids are still not well characterized and the mechanisms underlying the complex regulatory network of pathways have not been completely elucidated. Therefore, to achieve commercial production of more isoprenoids, the future work will include pathway characterization, enzyme screening and engineering, analysis and rebalancing of metabolic network, and development of nonconventional host strains. These processes will be greatly facilitated by the rapid development of synthetic biology tools.

References

1. Hunter, W.N. The non-mevalonate pathway of isoprenoid precursor biosynthesis. *J. Biol. Chem.* **282**, 21573–21577 (2007).
2. Maury, J., Asadollahi, M.A., Møller, K., Clark, A., & Nielsen, J. Microbial isoprenoid production: An example of green chemistry through metabolic engineering. *Adv. Biochem. Eng. Biotechnol.* **100**, 19–51 (2005).
3. Sandmann, G., Albrecht, M., Schnurr, G., Knörzer, O., & Böger, P. The biotechnological potential and design of novel carotenoids by gene combination in *Escherichia coli. Trends Biotechnol.* **17**, 233–237 (1999).
4. Roberts, S.C. Production and engineering of terpenoids in plant cell culture. *Nat. Chem. Biol.* **3**, 387–395 (2007).
5. Daum, M., Herrmann, S., Wilkinson, B., & Bechthold, A. Genes and enzymes involved in bacterial isoprenoid biosynthesis. *Curr. Opin. Chem. Biol.* **13**, 180–188 (2009).
6. Campobasso, N. *et al. Staphylococcus aureus* 3-hydroxy-3-methylglutaryl-CoA synthase crystal structure and mechanism. *J. Biol. Chem.* **279**, 44883–44888 (2004).
7. Anderson, M.S., Muehlbacher, M., Street, I., Proffitt, J., & Poulter, C.D. Isopentenyl diphosphate: Dimethylallyl diphosphate isomerase. An improved purification of the enzyme and isolation of the gene from *Saccharomyces cerevisiae. J. Biol. Chem.* **264**, 19169–19175 (1989).
8. Kaneda, K., Kuzuyama, T., Takagi, M., Hayakawa, Y., & Seto, H. An unusual isopentenyl diphosphate isomerase found in the mevalonate pathway gene

cluster from Streptomyces sp. strain CL190. *Proc. Natl. Acad. Sci. USA* **98**, 932–937 (2001).

9. Barkley, S.J., Cornish, R.M., & Poulter, C.D. Identification of an archaeal type II isopentenyl diphosphate isomerase in *Methanothermobacter thermautotrophicus. J. Bacteriol.* **186**, 1811–1817 (2004).

10. Dimster-Denk, D. *et al.* Comprehensive evaluation of isoprenoid biosynthesis regulation in Saccharomyces cerevisiae utilizing the Genome Reporter Matrix™. *J. Lipid Res.* **40**, 850–860 (1999).

11. Zhou, D., & White, R.H. Early steps of isoprenoid biosynthesis in *Escherichia coli. Biochem. J.* **273**, 627–634 (1991).

12. Flesch, G., & Rohmer, M. Prokaryotic hopanoids: The biosynthesis of the bacteriohopane skeleton. *Eur. J. Biochem.* **175**, 405–411 (1988).

13. Banerjee, A., & Sharkey, T. Methylerythritol 4-phosphate (MEP) pathway metabolic regulation. *Nat. Prod. Rep.* **31**, 1043–1055 (2014).

14. Immethun, C.M., Hoynes-O'Connor, A.G., Balassy, A., & Moon, T.S. Microbial production of isoprenoids enabled by synthetic biology. *Front Microbiol.* **4**, 75 (2013).

15. Guenther, A. *et al.* Estimates of global terrestrial isoprene emissions using MEGAN (Model of Emissions of Gases and Aerosols from Nature). *Atmos. Chem. Phys. Discuss.* **6**, 3181–3210 (2006).

16. Whited, G.M., Feher, F.J., & Benko, D.A. Development of a gas-phase bioprocess for isoprene-monomer production using metabolic pathway engineering. *Ind. Biotechnol.* **6**, 152–163 (2010).

17. Caputi, L., & Aprea, E. Use of terpenoids as natural flavouring compounds in food industry. *Recent Pat. on Food* **3**, 9–16 (2011).

18. Carter, O.A., Peters, R.J., & Croteau, R. Monoterpene biosynthesis pathway construction in *Escherichia coli. Phytochemistry* **64**, 425–433 (2003).

19. Munoz-Bertomeu, J., Ros, R., Arrillaga, I., & Segura, J. Expression of spearmint limonene synthase in transgenic spike lavender results in an altered monoterpene composition in developing leaves. *Metab. Eng.* **10**, 166–177 (2008).

20. Herrero, O., Ramon, D., & Orejas, M. Engineering the *Saccharomyces cerevisiae* isoprenoid pathway for de novo production of aromatic monoterpenes in wine. *Metab. Eng.* **10**, 78–86 (2008).

21. Rico, J., Pardo, E., & Orejas, M. Enhanced production of a plant monoterpene by overexpression of the 3-hydroxy-3-methylglutaryl coenzyme A reductase catalytic domain in *Saccharomyces cerevisiae. Appl. Environ. Microbiol.* **76**, 6449–6454 (2010).

22. Newman, R.D. *World Malaria Report 2011.* (2012).
23. Bouwmeester, H.J. *et al.* Amorpha-4,11-diene synthase catalyses the first probable step in artemisinin biosynthesis. *Phytochemistry* **52**, 843–854 (1999).
24. Bertea, C.M. *et al.* Identification of intermediates and enzymes involved in the early steps of artemisinin biosynthesis in *Artemisia annua. Planta Med.* **71**, 40–47 (2005).
25. Martin, V.J., Pitera, D.J., Withers, S.T., Newman, J.D., & Keasling, J.D. Engineering a mevalonate pathway in *Escherichia coli* for production of terpenoids. *Nat. Biotechnol.* **21**, 796–802 (2003).
26. Newman, J.D. *et al.* High-level production of amorpha-4, 11-diene in a two-phase partitioning bioreactor of metabolically engineered *Escherichia coli. Biotechnol. Bioeng.* **95**, 684–691 (2006).
27. Pitera, D.J., Paddon, C.J., Newman, J.D., & Keasling, J.D. Balancing a heterologous mevalonate pathway for improved isoprenoid production in *Escherichia coli. Metab. Eng.* **9**, 193–207 (2007).
28. Pfleger, B.F., Pitera, D.J., Smolke, C.D., & Keasling, J.D. Combinatorial engineering of intergenic regions in operons tunes expression of multiple genes. *Nat. Biotechnol.* **24**, 1027–1032 (2006).
29. Tsuruta, H. *et al.* High-level production of amorpha-4, 11-diene, a precursor of the antimalarial agent artemisinin, in *Escherichia coli. PLoS One* **4**, e4489 (2009).
30. Chang, M.C., Eachus, R.A., Trieu, W., Ro, D.-K., & Keasling, J.D. Engineering *Escherichia coli* for production of functionalized terpenoids using plant P450s. *Nat. Chem. Biol.* **3**, 274–277 (2007).
31. Ro, D.-K. *et al.* Production of the antimalarial drug precursor artemisinic acid in engineered yeast. *Nature* **440**, 940–943 (2006).
32. Lenihan, J.R., Tsuruta, H., Diola, D., Renninger, N.S., & Regentin, R. Developing an industrial artemisinic acid fermentation process to support the cost-effective production of antimalarial artemisinin-based combination therapies. *Biotechnol. Prog.* **24**, 1026–1032 (2008).
33. Westfall, P.J. *et al.* Production of amorphadiene in yeast, and its conversion to dihydroartemisinic acid, precursor to the antimalarial agent artemisinin. *Proc. Natl. Acad. Sci. USA* **109**, E111–E118 (2012).
34. Paddon, C.J. *et al.* High-level semi-synthetic production of the potent antimalarial artemisinin. *Nature* **496**, 528–532 (2013).
35. Malik, S. *et al.* Production of the anticancer drug in *Taxus baccata* suspension cultures: A review. *Process Biochem.* **46**, 23–34 (2011).

36. Ajikumar, P.K. *et al.* Isoprenoid pathway optimization for Taxol precursor overproduction in *Escherichia coli. Science* **330**, 70–74 (2010).

37. Howat, S. *et al.* Paclitaxel: Biosynthesis, production and future prospects. *J. Mol. Biol.* **31**, 242–245 (2014).

38. Huang, Q., Roessner, C.A., Croteau, R., & Scott, A.I. Engineering *Escherichia coli* for the synthesis of taxadiene, a key intermediate in the biosynthesis of taxol. *Bioorg. Med. Chem.* **9**, 2237–2242 (2001).

39. Engels, B., Dahm, P., & Jennewein, S. Metabolic engineering of taxadiene biosynthesis in yeast as a first step towards Taxol (Paclitaxel) production. *Metab. Eng.* **10**, 201–206 (2008).

40. Zhou, K., Qiao, K., Edgar, S., & Stephanopoulos, G. Distributing a metabolic pathway among a microbial consortium enhances production of natural products. *Nat. Biotechnol.* **33**, 377–383 (2015).

41. Ghimire, G.P., Lee, H.C., & Sohng, J.K. Improved squalene production via modulation of the methylerythritol 4-phosphate pathway and heterologous expression of genes from *Streptomyces peucetius* ATCC 27952 in *Escherichia coli. Appl. Environ. Microbiol.* **75**, 7291–7293 (2009).

42. Moses, T. *et al.* Combinatorial biosynthesis of sapogenins and saponins in *Saccharomyces cerevisiae* using a C-16α hydroxylase from *Bupleurum falcatum. Proc. Natl. Acad. Sci. USA* **111**, 1634–1639 (2014).

43. Mata-Gomez, L.C., Montanez, J.C., Mendez-Zavala, A., & Aguilar, C.N. Biotechnological production of carotenoids by yeasts: An overview. *Microb. Cell Fact.* **13**, 12 (2014).

44. Das, A. *et al.* An update on microbial carotenoid production: Application of recent metabolic engineering tools. *Appl. Microbiol. Biotechnol.* **77**, 505–512 (2007).

45. Lee, P.C., & Schmidt-Dannert, C. Metabolic engineering towards biotechnological production of carotenoids in microorganisms. *Appl. Microbiol. Biotechnol.* **60**, 1–11 (2002).

46. Farmer, W.R., & Liao, J.C. Precursor balancing for metabolic engineering of lycopene production in *Escherichia coli. Biotechnol. Prog.* **17**, 57–61 (2001).

47. Yuan, L.Z., Rouvière, P.E., LaRossa, R.A., & Suh, W. Chromosomal promoter replacement of the isoprenoid pathway for enhancing carotenoid production in *E. coli. Metab. Eng.* **8**, 79–90 (2006).

48. Jing, Z. *et al.* Engineering central metabolic modules of *Escherichia coli* for improving β-carotene production. *Metab. Eng.* **17**, 42–50 (2013).

49. Xie, W., Ye, L., Lv, X., Xu, H., & Yu, H. Sequential control of biosynthetic pathways for balanced utilization of metabolic intermediates in *Saccharomyces cerevisiae*. *Metab. Eng.* **28**, 8–18 (2014).
50. Bai, C. *et al.* Exploiting a precise design of universal synthetic modular regulatory elements to unlock the microbial natural products in *Streptomyces*. *Proc. Natl. Acad. Sci. USA.* **112**, E90–E97 (2015).

Chapter 8

Xylose Metabolism and Its Metabolic Engineering Applications

*Maria K. McClintock and Kechun Zhang**

Department of Chemical Engineering and Materials Science,
University of Minnesota, Minneapolis, MN 55455-0431, USA
*kzhang@umn.edu

8.1. Introduction

Xylose is a five-carbon hemicellulosic sugar, commonly found in the cell wall of plants. It represents a large percentage of lignocellulosic feedstock (approximately one-third) and consequently is one of the major targets for fermentative utilization.[1] Thus far, microbial utilization of D-xylose is yet to be commercially viable. While a large number of bacterial and fungal species can use xylose for direct fermentation, there is an inverse correlation of rate to yield (i.e., species with high rates give low product yield) when looking at natural fermentation. With the advance of metabolic engineering efforts (Section 8.3), this correlation has been improved over the past 40 years.

Lignocellulosic biomass is considered a promising source for renewable energy and chemical products. Agricultural lands in the United States can provide one billion dry tons of biomass in addition to the biomass that can be sustainably generated from forest land and the reuse of materials from the logging industry.[2] Furthermore, D-xylose can be isolated from biomass with better efficiency and at milder conditions than glucose.[3] Lastly, utilization of biomass feedstock is necessary to reduce its competition with food products and for maximal renewability of biomass-based products.[4]

8.2. The Fundamentals of Xylose Metabolism in Microbes

In the past, there have been numerous reviews and research articles written about the metabolism of xylose in bacteria and fungi, with a large focus on the industrially relevant strains of *Saccharomyces cerevisiae* and *Escherichia coli*.[1,3,5–12] This section aims to provide a context for many metabolic engineering efforts that will be discussed later in the text.

8.2.1. *Transport into the Cell*

Transportation into and out of cells is one of the limiting factors for D-xylose utilization. Besides diffusion, this transportation is enabled by carrier proteins. These vary considerably between bacterial and fungal species, increasing the complexity of selecting an ideal carrier protein and host for D-xylose conversion. Many bacteria, including *E. coli*, tend to use active transport for D-xylose. In yeasts and fungi, transportation can happen by facilitated diffusion or active transport.

Bacteria, such as *E. coli, Clostridia, Lactococci*, and *Bacilli*, use active transport for uptake of D-xylose into the cell.[13] Many species contain several routes for D-xylose uptake: usually a high- and a low-affinity transporter. In *E. coli*, the high-affinity transporter (*XylFGH*) belongs to the ATP-binding cassette (ABC) family[14] (Figure 8.1). The low-affinity transporter (*XlyE*), on the contrary, belongs to the major facilitator super-family (MFS) and acts as a proton symporter (Figure 8.1), making it pH dependent.[1,15] These transporters have been shown to be slightly promiscuous by allowing transportation of arabinose, albeit at lower efficiencies.[16] Additionally, at least two mechanisms of repression have been reported for the uptake of D-xylose. These include the presence of arabinose or CRP-dependent control of *xyl* genes, both of which lead to the preferential substrate being consumed first.[16,17] Homologous transporters to *XylFGH* have been reported in *Clostridiales, Thermoanaerobacterales*, and *Bacillales* species; and other MFS transporters have been reported in

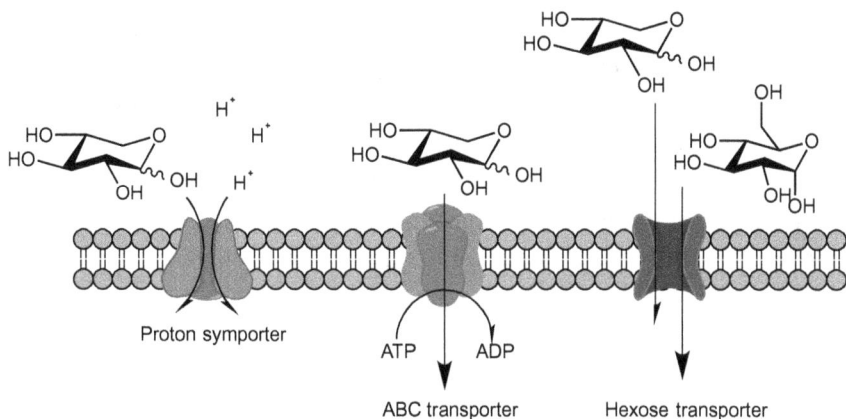

Figure 8.1.   Cartoon representations of different microbial transporter systems.

Bacilli and *Clostridia*.[13] Beyond these two types of transporters, phospho-transferase system (PTS) transporters have been reported in *Enterococcus faecalis* and *Clostridium difficile* for the uptake of the D-xylose precursor, xyloside.[13]

The transport and metabolism of D-xylose in yeasts and fungi is highly dependent on species and whether or not it can be utilized to sustain growth. Yeast, such as *Pichia*, *Candida*, and *Debaryomyces*, all have been reported with xylose-proton symporters that are substrate specific, have a high affinity, and are reliant upon induction by D-xylose for some species.[18–20] For other species, for example, *P. heedii*, glucose is a possible inhibitor of xylose transport similar to the previously mentioned inhibition in *E. coli*.[18] In addition to substrate-specific symporters, many yeasts that naturally metabolize D-xylose have symporters capable of transporting multiple sugars, such as glucose and D-xylose.[11] Unfortunately, the various studied transporters have a couple of drawbacks: (i) when utilized for multiple substrates a considerably lower affinity is reported for xylose than for glucose and (ii) some transporters have been found to be influenced by concentration of oxygen.[11,21] In addition to symporters, it is reported that many native consumers of D-xylose can uptake the substrate via facilitated diffusion.[18,20,22]

Saccharomyces cerevisiae, the industrial workhorse, cannot natively sustain cell growth on D-xylose. Naturally, *S. cerevisiae* relies predominately on facilitated diffusion through hexose transporters (HXT)[6] (Figure 8.1). Although the HXT can be divided into high- and low-affinity transporters for glucose (active under low and high concentrations of glucose, respectively), the affinities for xylose can be up to 200-fold less than that of glucose.[6,23] The lack of dedicated xylose transporters can potentially be the rate-limiting step depending on the engineered pathway. This is indicative of the need for carbon flux balancing in engineered *S. cerevisiae* (Section 8.3).[24]

8.2.2.   *Xylose Metabolism Through the Pentose Phosphate Pathway*

Once D-xylose has been consumed, there are three possible routes for utilization. The first two possibilities are (i) an isomerase converts the D-xylose into D-xylulose; or (ii) a reductase converts D-xylose into xylitol and a dehydrogenase converts the xylitol into D-xylulose [Figure 8.2(a)]. After the D-xylose has been converted to D-xylulose, it is phosphorylated to D-xylulose phosphate, which is an intermediate for the pentose phosphate pathway (PPP). [This chapter will cover only the enzymes until xylulose in detail, as the PPP is used for many applications (see Refs. [3], [6], [11], [12], and [25]–[28] for additional information.)] The other method for incorporation is a non-phosphorylative pathway, which completely bypasses PPP and provides intermediates for other uses in the cell (Section 8.2.3).

8.2.2.1 *Conversion by isomerization*

Xylose isomerase (XI) is the enzyme responsible for the reversible reaction converting D-xylose to D-xylulose [Figure 8.2(a)]. XI is present in many prokaryotes and some eukaryotes, like plants and fungi.[29–31] Amino acid sequences of these groups cluster into two subgroups with high homology within the two groups, but lower homology (approximately 25%) between the groups.[30] Group one consists of species from Proteobacteria, Bacteroidetes, Thermotogae, Clostridia, Bacillales, and Plantae, while the other is composed of Deinococcus-Thermus and Actinobacteria.[32] The primary difference in XI from these groups is thermostability.[32] The second group tends to be more stable at higher

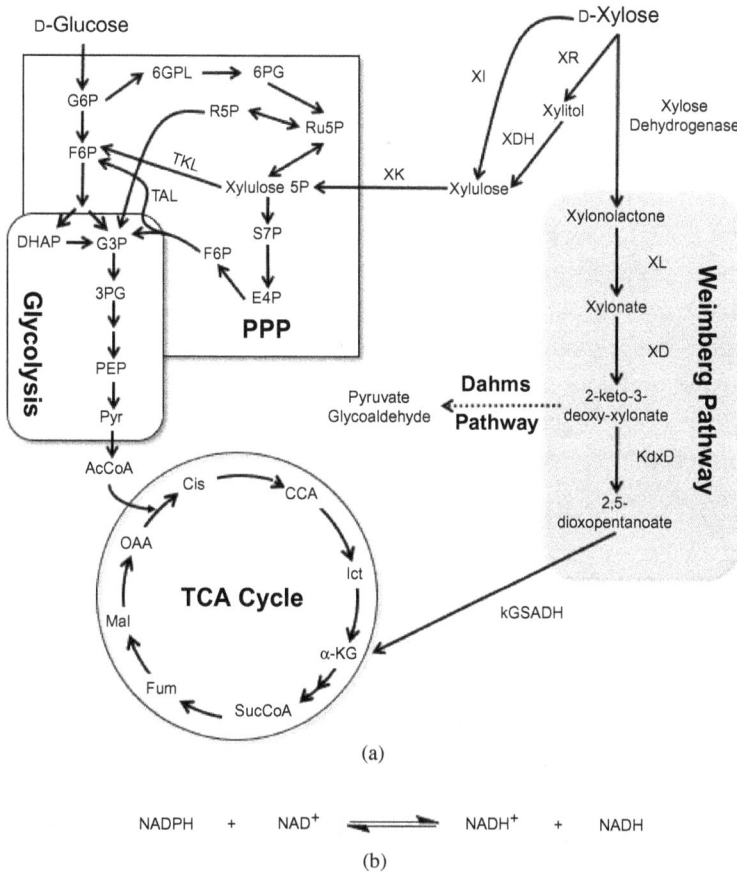

Figure 8.2. Overview of D-xylose metabolism: (a) shows the multiple routes for metabolism of D-xylose in microbes. For simplicity, only the genes discussed are labeled. Enzyme abbreviations are as follows: XI, D-xylose isomerase; XR, D-xylose reductase; XDH, D-xylitol dehydrogenase; XL, D-xylonolactonase; XD, D-xylonate dehydratase; KdxD, 2-keto-3-deoxy-D-xylonate dehydratase; KGSADH, 2-ketoglutarate semialdehyde dehydrogenase; XK, xylulokinase; TKL, transketolase; and TAL, transaldolase; (b) shows the conversion of NADPH by transhydrogenases.

temperatures and has an overall higher GC content in the amino acid sequence than the first group.

XI varies in structure, regulation, and affinity based on the origin species. Overall, XI has an optimum temperature range from 50°C to 80°C and an optimum pH between 7 and 9; however, this is variable (e.g., *Lactobacillus brevis* has an optimum pH between 6 and 7).[30,33] Regulation also varies by species. In some, such as *E. coli,* the gene *xylR* can serve as a repressor, unless xylose is present, which then makes it an activator. In other bacteria genera, such as *Bacillus* and *Staphylococcus*, it is thought that only inhibition occurs.[30] Both stability and regulation are reliant on divalent cofactors. The primary cofactors are Mg^{2+}, Co^{2+}, or Mn^{2+}.[26] All have been reported to serve as activators and Co^{2+} has been reported to stabilize enzyme conformation. Use of divalent cations is advantageous, as XI does not rely on pyridine nucleotide-derived cofactors, such as NAD and NADP, and does not require regeneration.[32] While dependent on these specific cation cofactors, XI is also subject to inhibition by others, including Ca^{2+}.

8.2.2.2. *Conversion by reductase/dehydrogenase*

While XI is commonly used by prokaryotes, most yeasts and fungi capable of D-xylose metabolism utilize a two-step, reduction/oxidation to convert D-xylose into D-xylulose [Figure 8.2(a)].[34–36] Xylose reductase (XR) reduces D-xylose into xylitol utilizing NADPH as a reducing agent.[34] Xylitol is then oxidized to D-xylulose by means of a xylitol dehydrogenase (XDH), which consumes NAD^+ for the reaction.[34] D-xylulose is subsequently phosphorylated to D-xylulose phosphate and used in the PPP.

The preference of XR for NADPH, while XDH prefers NAD^+, creates a tendency for cofactor imbalance, leading to either reduced product or excretion of xylitol.[35] While this may be advantageous for the production of xylitol, it diminishes yields and productivity for ethanol and other downstream products.[37,38] In fact, the impact of a cofactor imbalance has been the subject of numerous research publications and review articles.[39] When oxygen is readily available, cells can utilize respiration to convert NADH back to NAD^+.[40] Under oxygen-limited conditions, however, this is a problem.[41] In addition, many fungi, including *S. cerevisiae,* are unable to use transhydrogenases, which regenerate NAD^+ and $NADPH$[41] [Figure 8.2(b)].

Yeasts, such as *Pachysolen tannophilus, Pichia stipitis (Scheffer-somyces stipitis), Candida shehatae,* and *Candida parapsilosis,* have found other means of compensating for this imbalance.[34,41,42] For example, *S. stipitis,* has two dehydrogenases that catalyze the reaction from xylitol to xylulose. One *xyl2* is part of the medium chain dehydrogenase/reductase superfamily and is related to sorbitol dehydrogenases, while the other is a short chain dehydrogenase.[43] Another example is *C. parapsilosis,* whose XR shows a higher affinity for NADH than NADPH, due to a natural mutation in the binding pocket from lysine to arginine at site 274.[41]

XR is typically a tightly bound, non-cooperative dimer whose subunits and are between 33 and 40 kDa.[41,44] XR is a part of the aldoketo reductase superfamily, and its preference for NADH or NADPH has been explored in a number of characterization and mutagenesis studies.[41,45–49] As mentioned above, XR has a binding pocket that is important for NADH/NADPH preference. In *P. stipitis,* it has been shown through site-directed mutagenesis that the lysine at site 270 is important for binding to the 2′ phosphate group of NADPH.[45] Similarly, using computational design to improve XR from *Candida boidinii,* a negatively charged binding pocket helps binding to NADH, which when compared to NADPH is missing the negatively charged 2′-phosphate.[46] This binding site in XR has been shown to be highly flexible due to $(\beta/\alpha)_8$ barrel fold, which makes the protein engineering of NADH preference complex.[49,50]

XDH, which catalyzes the reversible reaction from xylitol to xylulose, was first kinetically described in the 1980s.[51,52] XDH varies greatly by species, as some species can contain multiple different XDH with different affinities.[43] XDH belongs to the medium-chain dehydrogenase reductase superfamily and can be composed of two or more subunits depending on the species. XDH contains the classical features of this group: a Rossmann fold and a Zn^{2+} at the catalytic site, which is necessary for activity.[50,53,54] At the catalytic core, XDH contains an aspartate residue, which helps to bind the 2′- and 3′-hydroxyl groups in NAD^+.[54] Mutations that make the binding pocket more amenable to NADP+ are thermally destabilizing, but can be compensated for with additional sites for zinc binding.[55]

Despite the need for metabolic balancing, the use of an oxidoreductase system in engineering applications may be justified by reports that this pathway confers an advantage when growing strains on lignocellulosic hydrolysates.[11,54] This is because it may help to detoxify furfurals, which are harmful cytotoxic side products of the hydrolysis process.[56] However, further research of XR/XDH benefits over XI need to be done as *S. cerevisiae* has recently been shown to utilize the XI pathway while growing on lignocellulosic hydrolases that were not detoxified.[57]

8.2.3. *Non-Phosphorylative Metabolism*

Several organisms including, but not limited to, *Pseudomonasfragi*, *Burkholderiaxenovorans*, and *Caulobacter crescentus* exhibit a completely different method for D-xylose assimilation.[58–61] These organisms utilize a non-phosphorylative pathway (Figure 8.2) dependent upon NAD⁺, which is different from traditional metabolism, where substrates travel through the PPP and glycolysis into the TCA cycle. In non-phosphorylative metabolism, D-xylose is converted in approximately three steps into 2-keto-3-deoxy-xylonate. 2-Keto-3-deoxy-xylonate is either converted through two reactions to α-ketoglutarate, which is an intermediate for the TCA cycle, or consumed via the Dahms pathway to produce glycoaldehyde and pyruvate[60,61] [Figure 8.2(a)].

Non-phosphorylative metabolism was first reported in the 1960s in *Pseudomonas fragi*.[60,62] The first step in the reaction pathway utilizes a xylose dehydrogenase (note: this is not XDH, which is xylitol dehydrogenase) to convert D-xylose into D-xylonolactone and produces NADH. D-Xylonolactonase is then used to convert the D-xylonolactone to D-xylonate. In the third step before the pathway diverges, the D-xylonate is converted to 2-keto-3-deoxy-xylonate by a D-xylonate dehydratase.[58,59] The first path possible for the 2-keto-3-deoxy-xylonate involves two steps: (i) first it is converted to 2,5-dioxopentanoate or α-ketoglutarate semialdehyde using a 2-keto-3-deoxy-D-xylonate dehydratase and (ii) then 2,5-dioxopentanoate is converted to α-ketoglutarate (or 2-ketoglutarate) producing NADH.[58,60] The α-ketoglutarate is a key intermediate for the TCA and can be utilized for primary metabolism pathways. The second path possible for 2-keto-3-deoxy-xylonate was first described by Dahms in 1974 in *Pseudomonas.*[61] In this chapter, an aldolase was reported to convert the 2-keto-3-deoxy-xylonate to pyruvate and glycoaldehyde.

8.3. Metabolic Engineering Approaches

Over the past 40 years, there has been significant investment in the conversion of D-xylose to mainly ethanol and xylitol.[63,64] However, these are only two examples of the potential products, as PPP/glycolysis and the TCA cycle serve as the "metabolic hub of the cell"[58] [Figure 8.2(a)]. This means that many high-value products can be obtained, including amino acids, industrial chemicals, and plastic alternatives (e.g., ethylene glycol, butanol, 1,4-butandiol, and polyhydroxybutyrate).[58,65–67] (Figure 8.3) Many of these pathways require metabolic engineering and optimization for an efficient conversion of substrate into the desired products. In addition to the identification and characterization of new pathways for xylose metabolism, heterologous expression of genes in workhorse organisms, for example, *E. coli* and *S. cerevisiae*, continues to be explored. Heterologous expression must also be paired with the optimization of DNA sequence and proper regulation of protein expression for efficient conversion. Protein engineering, through any of the typical techniques (i.e., random mutagenesis, rational design), has also been used for the optimization of xylose transporters and the enzymes responsible for the conversion to D-xylulose. Furthermore, researchers have explored strain optimization, coutilization of sugars, and carbon flux analysis for the enhancement of titers, in an effort to make production economically viable.

Since the utilization of D-xylose and other lignocellulosic sugars can enable the development of so many products from renewable and sustainable resources, there are a huge number of published papers on this topic. A literature search of "engineering xylose metabolism" returns almost 2,500 papers in the last six months and almost 35,000 papers total. This list does not even include the papers that are tangentially related to this area and can be used for its development. In the following text, only a few recent efforts from heterologous expression, protein engineering,

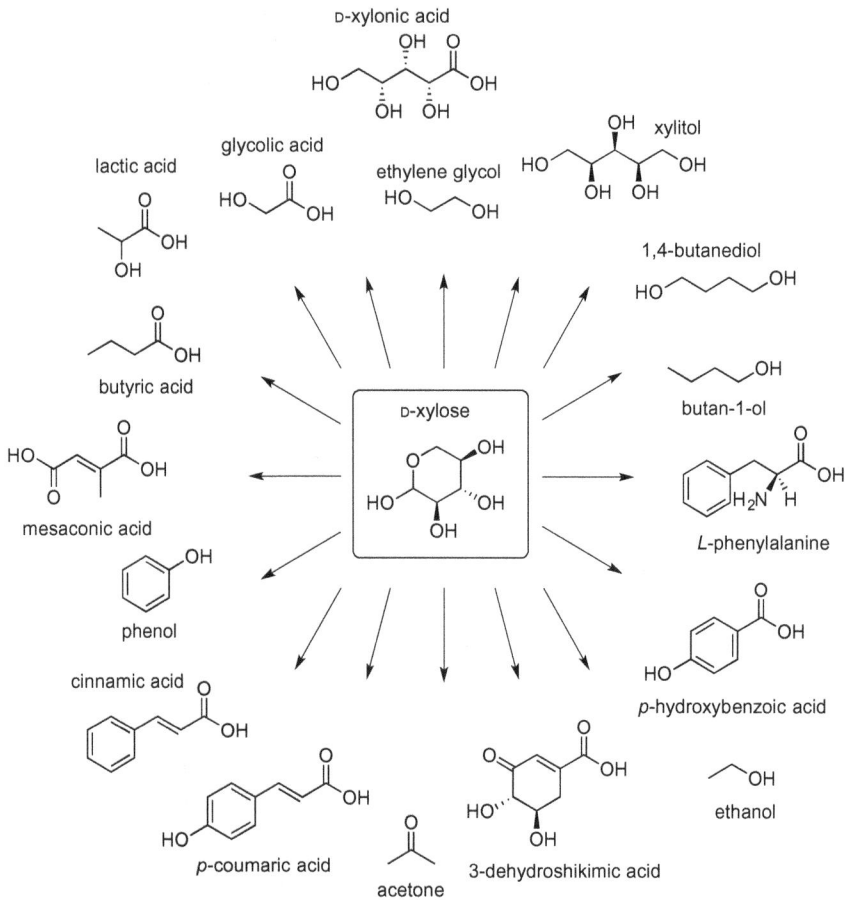

Figure 8.3. Exemplary industrially chemicals can be derived from metabolic engineering of microbes for D-xylose utilization. The chemicals shown above highlight the diversity of products and represent the efforts cited in this chapter.

carbon-flux analysis, and co-utilization will be discussed to highlight the variety of efforts.

8.3.1. *Heterologous Introduction of Genes*

E. coli and *S. cerevisiae* are the two most investigated species for xylose conversion. *E. coli* has the benefit of being a very thoroughly studied and characterized host that can naturally metabolize D-xylose. Unfortunately, *E. coli* does not produce high yields of ethanol and thus requires engineering to optimize production. *S. cerevisiae* is also well studied and well characterized and has the added benefit of being resilient to industrial production for chemicals, such as ethanol. However, *S. cerevisiae* cannot naturally utilize D-xylose to sustain growth and requires the addition of genes for xylose transport and conversion to D-xylulose.

8.3.1.1 *Examples of heterologous transporters*

As previously mentioned, *S. cerevisiae* cannot natively metabolize D-xylose, yet it is capable of fermentation of xylulose.[68] One of the reasons recombinant metabolism is difficult is because *S. cerevisiae* does not natively have any transporters dedicated to the uptake of D-xylose. A large number of *S. cerevisiae* monosaccharide transporters exist (HXR1-17, GAL2), but these transporters, as mentioned, have a higher affinity for hexose sugars, making the rate of uptake too low to sustain growth.[69] One potential remedy is to engineer *S. cerevisiae* by adding heterologous sugar transporters. Genes from several species, including *Candida intermedia, P. stipitis,* and *Arabidopsis thaliana*, have been used for improvement of utilization and titer.[69–72] While many of these attempts have been successful, the working concentration that shows the greatest change in growth rate is still low (i.e., 4 g/L, aerobic).[69] In addition to using heterologous genes, it is also important to address xylose transporter expression and control.[73,74] Transport is an important bottleneck in both *E. coli and S. cerevisiae*, which needs to be addressed for the utilization of D-xylose; however, this also must be matched with a balanced metabolic pathway.[75]

8.3.1.2 *Heterologous isomerases*

To effectively incorporate the conversion of D-xylose to D-xylulose in the PPP, many researchers examine heterologous expression of XI in both *E. coli* and yeast as a promising route. XI have been found in many organisms (see Section 8.2.2.1), leading to a large array of protein space that can be explored.[26,27,29,38,57,65,76–80] There are several features of recombinant isomerase that are important for industrial usage, depending on whether the application is *in vitro* or *in vivo*. One such feature is activity at a lower pH. This feature is particularly important for yeast fermentations, which are conducted at a pH from 4 to 5.[78] Most XIs are optimum at a pH 7, but an XI recombinantly expressed in *E. coli* from *Lactobacillus reuteri* has been shown to have optimal activity at a pH of 5.5.[78,81] A second desirable feature is high resistance to xylitol and other potential inhibitors that might come from either hydrolysis of biomass or from the fermentation conditions. These characteristics have been reported in an XI isolated from *Fulvimarina pelagi*, which was expressed in *E. coli*.[78] These are only two examples of features to look for when identifying the appropriate XI for a new system.

8.3.1.3 *Heterologous reductase/dehydrogenase systems*

Similar to XI, many heterologous reductase/dehydrogenase systems have been explored for use in *E. coli* and *S. cerevisiae*. In *E. coli, Corynebacterium glutamicum*, and *Lactococcus lactis*, the usage of XR or XDH is predominately for the conversion of xylitol, which is an intermediate of the XR/XDH system.[39,82–87] XR from *Candida tenuis* has also been used in an engineered pathway for production of ethyl R-4-cyanomandelate in both *E. coli* and *S. cerevisiae*.[88] As a system, XR and XDH are predominately explored for ethanol production in *S. cerevisiae* due to its ability to afford high ethanol production.[10,42,47,89–93] Additionally, many of these systems have been optimized through strain development or protein engineering, which will be discussed in Section 8.3.2. For all recombinant use of XR and XDH, there needs to be careful examination of the metabolic burden placed upon the cell due to the usage of NADPH/NADP+ and NADH/NAD+.[84]

8.3.1.4 *Heterologous usage of downstream pathways*

Beyond engineering the enzymes that feed into the PPP, it is often necessary to upregulate or optimize the PPP enzymes themselves. Xylulokinase, responsible for the conversion of xylulose to xylulose-5-phosphate through the consumption of ATP, is one such target for upregulation and heterologous introduction.[25,27,28,81,93–95] One example of this is the overexpression of xylulokinase from *P. stipitis* in *S. cerevisiae*. This xylulokinase was shown to have 6-fold higher activity when on a multicopy vector.[94] This higher activity, however, seemingly caused a lower ethanol production. Researchers have also tried to upregulate the expression levels of TKL and TAL for improvement in xylose fermentation.[96] TKL and TAL are two genes in the non-oxidative portion of the PPP[25] (TKL and TAL, respectively, Figure 8.2). Overproduction of TAL only showed increased consumption of xylulose, not xylose.[68,97] Furthermore, significant changes in growth were seen only on xylulose and only together with the upregulation of ribulose-5-phosphate epimerase and ribose-5-phosphate ketol isomerase.[68] These works highlight the complexity of optimally engineering an essential pathway.

Several groups have also worked on heterologous or upregulated expression of genes for use outside of the PPP or the TCA cycle.[98] For example, Wang *et al.* upregulated the use of FucO (propanediol oxidoreductase) and found that it increased *E. coli* resistance to furfural. This enabled the creation of a strain that can survive growth in 15 mM furfural, which is relevant for biomass hydrolysates.[99,100] Another excellent example of metabolic engineering that is not directly related to PPP is the creation of an engineered *E. coli* strain capable of producing ethylene glycol and glycolate.[101] The authors utilize the natural XI for production of xylulose and then employ recombinant genes from *Pseudomona scichorii* for the four step conversions to each product.[101]

8.3.1.5 *Heterologous usage of non-phosphorylative pathways*

In addition to engineering aspects of traditional metabolism, the non-phosphorylative pathway provides a number of opportunities for researchers to develop new and interesting products. In particular, the ability to turn D-xylose into α-ketoglutarate, via the Weimberg pathway, allows for efficient production of TCA cycle derivatives, while not placing as much of a metabolic burden on the engineered cells and being more energy efficient than traditional pathways.[58,96,102] Tai *et al.* recently engineered *E. coli* to utilize non-phosphporylative metabolism pathways for D-xylose, L-arabinose, and D-galacturonate.[58] The pathway from *Caulobacter crescentus* for D-xylose metabolism was used to design and test a platform for the identification of additional pathways that are functional in *E. coli*.

This platform enabled the authors to identify a previously uncharacterized D-xylose pathway from *Burkholderiaxenovorans*. The identified pathway was characterized and found to contain a 2-keto-3-deoxy-D-xylonate dehydratase with a 9-fold higher k_{cat} than the corresponding gene in *C. crescentus*, emphasizing the importance of gene discovery. The pathways described in this chapter were used for the production of 1,4-butandiol, as proof of concept. Other researchers have used enzymes in this pathway, specifically xylose dehydrogenase (note: this is not XDH, which is xylitol dehydrogenase), for recombinant production of both ethylene glycol and D-xylonic acid.[66,103]

8.3.2 *Protein Engineering for Enzymes*

Heterologous expression of genes is an excellent way to establish pathways for production. Unfortunately, few heterologous expressed systems are economically viable without further protein and strain optimization (see Section 8.3.3). Several mutagenesis studies were discussed in the metabolism sections for the identification of active sites and the improvement of enzyme activity. A sample of the protein engineering papers that have either utilized a particularly interesting method for protein development, or made a significant impact on the functionality of the enzyme, will be highlighted in this section.

The functionality of XR's binding pocket has been well characterized using mutagenesis (see Section 8.2.2.1 for additional references).[104–106] In addition to site-directed mutagenesis, researchers have also used computational design for altering the cofactor specificity. In this study, the cofactor binding site was examined by comparing mutational studies for a change from NADPH to NADH across enzymes from various species.[46] (Readers are directed to Table 1 in Ref. [46] for more information on the different mutations made for cofactor switching.) The authors found that across different enzymes, there is no direct correlation between any of the obvious changes, such as volume, hydrophobicity, or charge. Despite this, they were able to identify 10 mutation combinations that were predicted to improve the binding of XR from *Candida boidinii* with one having 27-fold higher affinity for NADH.

XDH and XI have also been explored for protein engineering. One particular study engineered xylitol dehydrogenase to be both thermally stable and change the cofactor usage to $NADP^+$.[54] This was done through mutations to the binding site domain and the addition of an additional zinc atom to the catalytic domain.[54,55] [Readers are directed to Figure 1 of Ref. [54] and Figure 1 of Ref. [55] for more information.] Similarly, XI has been engineered for improved thermal stability. Site-directed mutagenesis was able to improve XI from *Thermotoga neapolitana*'s maximum activity to 97°C.[107] This higher temperature could be of great benefit in commercial enzymatic synthesis applications.

8.3.3 *Strain Optimization*

In order to create economically viable and sustainable processes, strain engineering must be used to (i) allow for optimal use of multiple sugars; (ii) create strains resilient to inhibitors present in hydrolysates; and (iii) optimally direct the carbon flow toward the desired product. Many of the studies covered in this chapter have conducted some strain optimization through gene deletions, gene upregulation, and through directed evolutions.

Utilization of multiple sugar substrates is important for the creation of industrially relevant renewable fermentation processes. For lignocellulosic feedstocks to be optimally exploited, all of the sugars present, including but not limited to arabinose, xylose, and glucose, must be used to sustain cell growth while also producing the product. Many different methods for optimal co-utilization have been explored.[24,95,98,101,108–112] Several computational tools have allowed for the production of industrially relevant chemicals via utilization of multiple sugar substrates. Yim *et al.* used a pathway identification algorithm to design a heterologous pathway to BDO that is capable of producing 20 g/L BDO.[112]

Strain optimization is another important step for the creation of industrially relevant species. Many heterologous pathway designs involve minor strain optimization through deletion of genes that would consume the final product (i.e., the elimination of genes that consume xylitol). Strain optimization studies are frequently used to eliminate bottlenecks.[28,113–115] Xiao *et al.* used strain optimization on *Clostridium acetobutylicum* and were able to create a strain capable of approximately 16 g/L of solvent (mixture of acetone, butanol, and ethanol) while using a mixture of glucose, xyloses, and arabinose as substrates.[28] Furthermore, carbon flux analysis of diverse hosts can inform their optimization, allow for the use of new hosts in industrial processes, and inform the design of pathways in heterologous hosts.[114]

8.4. Conclusions and Future Applications

This chapter aims to provide an overview of the metabolic engineering efforts made for the utilization of D-xylose and the required background description of the fundamentals of xylose metabolism in bacteria and fungi. Looking forward, there is much work to be done if xylose and other lignocellulosic sugars are to be utilized for renewable and sustainable production of fuels and chemicals.

Future developments of xylose metabolism can first look to the expansion of products derived from the TCA cycle. Based on the non-phosphorylative pathway, products such as 1,4-butanediol, 1,2,4-butanetriol, and glutamate can be produced utilizing the three main components of lignocellulosic feedstocks without needing to go through the PPP or the TCA cycle.[7,58] In addition, to expand the range of both potential products (Figure 8.3) and substrates, further developments will have to occur in metabolic engineering and in enzyme engineering. This includes, but is not limited to (i) strain optimization through carbon flux analysis and directed evolution; (ii) enzyme engineering to improve the efficacy through techniques such as directed evolution of proteins, rational design, and ancestral reconstruction; and (iii) use of synthetic biology techniques to optimize the transcriptional and translational regulation of genes in the host organism.

To date, most engineered systems focus on utilization of D-xylose in *E. coli* and *S. cerevisiae*. With these organisms, there is the potential for the coutilization of sugars that has been explored, but researchers can potentially look to other organisms that may be better or more efficient hosts, depending on the nature of the product and the industry it is used in. This, however, will require the development of additional tools for genetic engineering of other, non-standard organisms. In addition, there is currently no good method for the standardization to characterize xylose-related enzymes across organisms, making the selection and design of pathways more complex.

Furthermore, there are external developments that can occur. For example, advances in the hydrolysis of biomass, whether enzymatic or by acid, may play a role in the design of an industrially relevant species. Additionally, developing strains that are more resilient to the various other compounds, for example, furfural, which is found after acid hydrolysis of biomass, is another area with scope for development.[33,100] Finally, all of the genes and enzymes that have been identified or engineered for use in metabolism of D-xylose can potentially contribute to the conversion or consumption of other products and substrates, respectively.[88]

References

1. McMillan, J.D. *Xylose fermentation to ethanol: A review*, NREL, US Department of Energy, Golden, Colorado (1993).
2. Perlack, R.D. *et al. Biomass as feedstock for a bioenergy and bioproducts industry: The technical feasibility of a billion-ton annual supply*. Oak Ridge National Lab, US Department of Energy, Oak Ridge, TN (2005).
3. Jeffries, T.W. Utilization of xylose by bacteria, yeasts, and fungi. *Adv. Biochem. Eng. Biotechnol.* **27**, 1–32 (1983).
4. Graham-Rowe, D. Agriculture: Beyond food versus fuel. *Nature* **474**, S6–S8 (2011).
5. Akinterinwa, O., Khankal, R., & Cirino, P.C. Metabolic engineering for bioproduction of sugar alcohols. *Curr. Opin. Biotechnol.* **19**, 461–467 (2008).
6. Chu, B.C.H., & Lee, H. Genetic improvement of *Saccharomyces cerevisiae* for xylose fermentation. *Biotechnol. Adv.* **25**, 425–441 (2007).
7. Danner, H., & Braun, R. Biotechnology for the production of commodity chemicals from biomass. *Chem. Soc. Rev.* **28**, 395–405 (1999).
8. Geddes, C.C., Nieves, I.U., & Ingram, L.O. Advances in ethanol production. *Curr. Opin. Biotechnol.* **22**, 312–319 (2011).
9. McMillan, J.D. Bioethanol production: Status and prospects. *Renew. Energy* **10**, 295–302 (1997).

10. Jeffries, T.W. Engineering yeasts for xylose metabolism. *Curr. Opin. Biotechnol.* **17**, 320–326 (2006).

11. Van Vleet, J., & Jeffries, T.W. Yeast metabolic engineering for hemicellulosic ethanol production. *Curr. Opin. Biotechnol.* **20**, 300–306 (2009).

12. Skoog, K., & Hahn-Hagerdal, B. Xylose fermentation. *Enzyme Microb. Technol.* **10**, 66–80 (1988).

13. Gu, Y. *et al.* Reconstruction of xylose utilization pathway and regulons in *Firmicutes. BMC Genomics* **11**, 255–269 (2010).

14. Lee, T.-C. *et al.* Engineered xylose utilization enhances bio-products productivity in the cyanobacterium *Synechocystis* sp. PCC 6803. *Metab. Eng.* **30**, 179–189 (2015).

15. Sumiya, M., Davis, E.O., Packman, L.C., McDonald, T.P., & Henderson, P.J. Molecular genetics of a receptor protein for D-xylose, encoded by the gene xylF, in *Escherichia coli. Recept. Channel.* **3**, 117–128 (1995).

16. Desai, T.A., & Rao, C.V. Regulation of arabinose and xylose metabolism in *Escherichia coli. Appl. Environ. Microbiol.* **76**, 1524–1532 (2010).

17. Cirino, P.C., Chin, J.W., & Ingram, L.O. Engineering *Escherichia coli* for xylitol production from glucose-xylose mixtures. *Biotechnol. Bioeng.* **95**, 1167–1176 (2006).

18. Does, A.L., & Bisson, L.F. Characterization of xylose uptake in the yeasts *Pichia heedii* and *Pichia stipitis. Appl. Environ. Microbiol.* **55**, 159–164 (1989).

19. Kilian, S.G., Prior, B.A., & du Preez, J.C. The kinetics and regulation of D-xylose transport in *Candida utilis. World J. Microbiol. Biotechnol.* **9**, 357–360 (1993).

20. Nobre, A., Lucas, C., & Leao, C. Transport and utilization of hexoses and pentoses in the halotolerant yeast *Debaryomyces hansenii. Appl. Environ. Microbiol.* **65**, 3594–3598 (1999).

21. Ligthelm, M.E., Prior, B.A., du Preez, J.C., & Brandt, V. An investigation of D-(1-^{13}C) xylose metabolism in *Pichia stipitis* under aerobic and anaerobic conditions. *Appl. Microbiol. Biotechnol.* **28**, 293–296 (1988).

22. Kilian, S.G., & van Uden, N. Transport of xylose and glucose in the xylose-fermenting yeast *Pichia stipitis. Appl. Microbiol. Biotechnol.* **27**, 545–548 (1988).

23. Kotter, P., & Ciriacy, M. Xylose fermentation by *Saccharomyces cerevisiae. Appl. Microbiol. Biotechnol.* **38**, 776–783 (1993).

24. Reider Apel, A., Ouellet, M., Szmidt-Middleton, H., Keasling, J.D., & Mukhopadhyay, A. Evolved hexose transporter enhances xylose uptake and glucose/xylose co-utilization in *Saccharomyces cerevisiae*. *Sci. Rep.* **6**, 1–10 (2016).
25. Johansson, B., Christensson, C., Hobley, T., & Hahn-Hagerdal, B. Xylulokinase overexpression in two strains of *Saccharomyces cerevisiae* also expressing xylose reductase and xylitol dehydrogenase and its effect on fermentation of xylose and lignocellulosic hydrolysate. *Appl. Environ. Microbiol.* **67**, 4249–4255 (2001).
26. Zheng, Z., Lin, X., Jiang, T., Ye, W., & Ouyang, J. Genomic analysis of a xylose operon and characterization of novel xylose isomerase and xylulokinase from *Bacillus coagulans NL01*. *Biotechnol. Lett.* **38**(8), 1331–1339 (2016).
27. Dmytruk, O.V. *et al.* Overexpression of bacterial xylose isomerase and yeast host xylulokinase improves xylose alcoholic fermentation in the thermotolerant yeast *Hansenula polymorpha*. *FEMS Yeast Res.* **8**, 165–173 (2008).
28. Xiao, H. *et al.* Confirmation and elimination of xylose metabolism bottlenecks in glucose phosphoenolpyruvate-dependent phosphotransferase system-deficient *Clostridium acetobutylicum* for simultaneous utilization of glucose, xylose, and arabinose. *Appl. Environ. Microbiol.* **77**, 7886–7895 (2011).
29. Madhavan, A. *et al.* Xylose isomerase from polycentric fungus *Orpinomyces*: Gene sequencing, cloning, and expression in *Saccharomyces cerevisiae* for bioconversion of xylose to ethanol. *Appl. Microbiol. Biotechnol.* **82**, 1067–1078 (2009).
30. Bhosale, S.H., Rao, M.B., & Deshpande, V.V. Molecular and industrial aspects of glucose isomerase. *Microbiol. Rev.* **60**, 280–300 (1996).
31. Riveros-Rosas, H., Julián-Sánchez, A., Villalobos-Molina, R., Pardo, J.P., & Piña, E. Diversity, taxonomy and evolution of medium-chain dehydrogenase/reductase superfamily. *Eur. J. Biochem.* **270**, 3309–3334 (2003).
32. van Maris, A.J.A. *et al.* Development of efficient xylose fermentation in *Saccharomyces cerevisiae*: Xylose isomerase as a key component. *Adv. Biochem. Eng. Biotechnol.* **108**, 179–204 (2007).
33. Olsson, L., & Hahn-Hagerdal, B. Fermentation of lignocellulosic hydrolysates for ethanol production. *Enzyme Microb. Technol.* **18**, 312–331 (1996).

34. Bruinenberg, P.M., de Bot, P.H.M., van Dijken, J.P., & Scheffers, W.A. The role of redox balances in the anaerobic fermentation of xylose by yeasts. *Eur. J. Appl. Microbiol. Biotechnol.* **18**, 287–292 (1983).

35. Bruinenberg, P.M., de Bot, P.H.M., van Dijken, J.P., & Scheffers, W.A. NADH-linked aldose reductase: The key to anaerobic alcoholic fermentation of xylose by yeasts. *Appl. Microbiol. Biotechnol.* **19**, 256–260 (1984).

36. Zhang, M. *et al.* Genetic analysis of D-xylose metabolism pathways in *Gluconobacter oxydans* 621H. *J. Ind. Microbiol. Biotechnol.* **40**, 379–388 (2013).

37. Girio, F.M., Peito, M.A., & Amaral-Collaco, M.T. Enzymatic and physiological study of D-xylose metabolism by *Candida shehatae*. *Appl. Microbiol. Biotechnol.* **32**, 199–204 (1989).

38. Zhou, H., Cheng, J.-S., Wang, B.L., Fink, G.R., & Stephanopoulos, G. Xylose isomerase overexpression along with engineering of the pentose phosphate pathway and evolutionary engineering enable rapid xylose utilization and ethanol production by *Saccharomyces cerevisiae*. *Metab. Eng.* **14**, 611–622 (2012).

39. Chin, J.W., & Cirino, P.C. Improved NADPH supply for xylitol production by engineered *Escherichia coli* with glycolytic mutations. *Biotechnol. Prog.* **27**, 333–341 (2011).

40. Jin, Y.-S., & Jeffries, T.W. Stoichiometric network constraints on xylose metabolism by recombinant *Saccharomyces cerevisiae*. *Metab. Eng.* **6**, 229–238 (2004).

41. Lee, J.-K., Koo, B.S., & Kim, S.Y. Cloning and characterization of the xyl1 gene, encoding an NADH-preferring xylose reductase from *Candida parapsilosis*, and its functional expression in *Candida tropicalis*. *Appl. Environ. Microbiol.* **69**, 6179–6188 (2003).

42. Hahn-Hagerdal, B., Jeppsson, H., Skoog, K., & Prior, B.A. Biochemistry and physiology of xylose fermentation by yeasts. *Enzyme Microb. Technol.* **16**, 933–943 (1994).

43. Persson, B. *et al.* Dual relationships of xylitol and alcohol dehydrogenases in families of two protein types. *FEBS Lett.* **324**, 9–14 (1993).

44. Yokoyama, S.I., Suzuki, T., Kawai, K., Horitsu, H., & Takamizawa, K. Purification, characterization and structure analysis of NADPH-dependent D-xylose reductases from *Candida tropicalis*. *J. Ferment. Bioeng.* **79**, 217–223 (1995).

45. Kostrzynska, M., Sopher, C.R., & Lee, H. Mutational analysis of the role of the conserved lysine-270 in the *Pichia stipitis* xylose reductase. *FEMS Microbiol. Lett.* **159**, 107–112 (1998).
46. Khoury, G.A. *et al.* Computational design of *Candida boidinii* xylose reductase for altered cofactor specificity. *Protein Sci.* **18**, 2125–2138 (2009).
47. Watanabe, S. *et al.* The positive effect of the decreased NADPH-preferring activity of xylose reductase from *Pichia stipitis* on ethanol production using xylose-fermenting recombinant *Saccharomyces cerevisiae. Biosci. Biotechnol. Biochem.* **71**, 1365–1369 (2007).
48. Watanabe, S. *et al.* Ethanol production from xylose by recombinant *Saccharomyces cerevisiae* expressing protein-engineered NADH-preferring xylose reductase from *Pichia stipitis. Microbiology* **153**, 3044–3054 (2007).
49. Kavanagh, K.L., Klimacek, M., Nidetzky, B., & Wilson, D.K. Structure of xylose reductase bound to NAD+ and the basis for single and dual co-substrate specificity in family 2 aldo-keto reductases. *Biochem. J.* **373**, 319–326 (2003).
50. Krahulec, S., Klimacek, M., & Nidetzky, B. Analysis and prediction of the physiological effects of altered coenzyme specificity in xylose reductase and xylitol dehydrogenase during xylose fermentation by *Saccharomyces cerevisiae. J. Biotechnol.* **158**, 192–202 (2012).
51. Rizzi, M., Harwart, K., Bui-Thanh, N.-A., & Dellweg, H. A kinetic study of the NAD⁺-xylitol-dehydrogenase from the yeast *Pichia stipitis. J. Ferment. Bioeng.* **67**, 25–30 (1989).
52. Rizzi, M., Harwart, K., Erlemann, P., Bui-Thanh, N.-A., & Dellweg, H. Purification and properties of the NAD⁺-xylitol-dehydrogenase from the yeast *Pichia stipitis. J. Ferment. Bioeng.* **67**, 20–24 (1989).
53. Lunzer, R., Mamnun, Y., Haltrich, D., Kulbe, K.D., & Nidetzky, B. Structural and functional properties of a yeast xylitol dehydrogenase, a Zn^{2+}-containing metalloenzyme similar to medium-chain sorbitol dehydrogenases. *Biochem. J.* **336**, 91–99 (1998).
54. Watanabe, S., Kodaki, T., & Makino, K. Complete reversal of coenzyme specificity of xylitol dehydrogenase and increase of thermostability by the introduction of structural zinc. *J. Biol. Chem.* **280**, 10340–10349 (2005).
55. Annaluru, N. *et al.* Thermostabilization of *Pichia stipitis* xylitol dehydrogenase by mutation of structural zinc-binding loop. *J. Biotechnol.* **129**, 717–722 (2007).

56. Almeida, J.R.M., Modig, T., Röder, A., Lidén, G., & Gorwa-Grauslund, M.F. *Pichia stipitis* xylose reductase helps detoxifying lignocellulosic hydrolysate by reducing 5-hydroxymethyl-furfural (HMF). *Biotechnol. Biofuels* **1**, 1–9 (2008).
57. Ko, J.K., Um, Y., Woo, H.M., Kim, K.H., & Lee, S.-M. Ethanol production from lignocellulosic hydrolysates using engineered *Saccharomyces cerevisiae* harboring xylose isomerase-based pathway. *Bioresour. Technol.* **209**, 290–296 (2016).
58. Tai, Y.-S. *et al.* Engineering nonphosphorylative metabolism to generate lignocellulose-derived products. *Nat. Chem. Biol.* **12**, 247–253 (2016).
59. Stephens, C. *et al.* Genetic analysis of a novel pathway for D-xylose metabolism in *Caulobacter crescentus*. *J. Bacteriol.* **189**, 2181–2185 (2007).
60. Weimberg, R. Pentose oxidation by *Pseudomonas fragi*. *J. Biol. Chem.* **236**, 629–635 (1961).
61. Dahms, A.S. 3-Deoxy-D-pentulosonic acid aldolase and its role in a new pathway of D-xylose degradation. *Biochem. Biophys. Res. Commun.* **60**, 1433–1439 (1974).
62. Poindexter, J.S. Biological properties and classification of the *Caulobacter* group. *Bacteriol. Rev.* **28**, 231–295 (1964).
63. Chin, J.W., & Cirino, P.C. In: P. Wang (ed.), *methods in molecular biology*. Humana Press, Totowa, NJ, Vol. 743, 185–203, (2011).
64. Hacker, B., Habenicht, A., Kiess, M., & Mattes, R. Xylose utilisation: Cloning and characterisation of the xylose reductase from *Candida tenuis*. *Biol. Chem.* **380**, 1395–1403 (1999).
65. Lopes, M.S.G., Gomez, J.G.C., & Silva, L.F. Cloning and overexpression of the xylose isomerase gene from *Burkholderia sacchari* and production of polyhydroxybutyrate from xylose. *Can. J. Microbiol.* **55**, 1012–1015 (2009).
66. Liu, H. *et al.* Biosynthesis of ethylene glycol in *Escherichia coli*. *Appl. Microbiol. Biotechnol.* **97**, 3409–3417 (2013).
67. Liu, H., & Lu, T. Autonomous production of 1,4-butanediol via a de novo biosynthesis pathway in engineered *Escherichia coli*. *Metab. Eng.* **29**, 135–141 (2015).
68. Johansson, B., & Hahn-Hagerdal, B. The non-oxidative pentose phosphate pathway controls the fermentation rate of xylulose but not of xylose in TMB3001. *FEMS Yeast Res.* **2**, 277–282 (2002).
69. Runquist, D., Hahn-Hägerdal, B., & Rådström, P. Comparison of heterologous xylose transporters in recombinant *Saccharomyces cerevisiae*. *Biotechnol. Biofuels* **3**, 1–7 (2010).

70. Runquist, D., Fonseca, C., Rådström, P., Spencer-Martins, I., & Hahn-Hägerdal, B. Expression of the Gxf1 transporter from *Candida intermedia* improves fermentation performance in recombinant xylose-utilizing *Saccharomyces cerevisiae*. *Appl. Microbiol. Biotechnol.* **82**, 123–130 (2009).

71. Leandro, M.J., Goncalves, P., & Isabel, S.-M. Two glucose/xylose transporter genes from the yeast *Candida intermedia*: First molecular characterization of a yeast xylose-H$^+$ symporter. *Biochem. J.* **395**, 543–549 (2006).

72. Weierstall, T., Hollenberg, C.P., & Boles, E. Cloning and characterization of three genes (SUT1-3) encoding glucose transporters of the yeast *Pichia stipitis*. *Mol. Microbiol.* **31**, 871–883 (1999).

73. Salusjärvi, L. *et al.* Regulation of xylose metabolism in recombinant *Saccharomyces cerevisiae*. *Microb. Cell Fact.* **7**, 1–16 (2008).

74. Khankal, R., Chin, J.W., & Cirino, P.C. Role of xylose transporters in xylitol production from engineered *Escherichia coli*. *J. Biotechnol.* **134**, 246–252 (2008).

75. Gardonyi, M., Jeppsson, M., Liden, G., Gorwa-Grauslund, M.F., & Hahn-Hagerdal, B. Control of xylose consumption by xylose transport in recombinant *Saccharomyces cerevisiae*. *Biotechnol. Bioeng.* **82**, 818–824 (2003).

76. Brat, D., Boles, E., & Wiedemann, B. Functional expression of a bacterial xylose isomerase in *Saccharomyces cerevisiae*. *Appl. Environ. Microbiol.* **75**, 2304–2311 (2009).

77. Parachin, N.S., & Gorwa-Grauslund, M.F. Isolation of xylose isomerases by sequence- and function-based screening from a soil metagenomic library. *Biotechnol. Biofuels* **4**, 9 (2011).

78. Lajoie, C.A. *et al.* Cloning, expression and characterization of xylose isomerase from the marine bacterium *Fulvimarina pelagi* in *Escherichia coli*. *Biotechnol. Prog.* **32**, 1230–1237 (2016).

79. de Figueiredo Vilela, L. *et al.* Functional expression of *Burkholderia cenocepacia* xylose isomerase in yeast increases ethanol production from a glucose-xylose blend. *Bioresour. Technol.* **128**, 792–796 (2013).

80. Vieilee, C., Hess, J.M., Kelly, R.M., & Zeikus, J.G. xylA cloning and sequencing and biochemical characterization of xylose isomerase from *Thermotoga neapolitana*. *Appl. Environ. Microbiol.* **61**, 1867–1875 (1995).

81. Staudigl, P., Haltrich, D., & Peterbauer, C.K. L-arabinose isomerase and D-xylose isomerase from *Lactobacillus reuteri*: Characterization, coexpression in the food grade host *Lactobacillus plantarum*, and application in the

conversion of D-galactose and D-glucose. *J. Agric. Food Chem.* **62**, 1617–1624 (2014).

82. Akinterinwa, O., & Cirino, P.C. Heterologous expression of D-xylulokinase from *Pichia stipitis* enables high levels of xylitol production by engineered *Escherichia coli* growing on xylose. *Metab. Eng.* **11**, 48–55 (2009).

83. Khankal, R., Luziatelli, F., Chin, J.W., Frei, C.S., & Cirino, P.C. Comparison between *Escherichia coli* K-12 strains W3110 and MG1655 and wild-type *E. coli* B as platforms for xylitol production. *Biotechnol. Lett.* **30**, 1645–1653 (2008).

84. Chin, J.W., Khankal, R., Monroe, C.A., Maranas, C.D., & Cirino, P.C. Analysis of NADPH supply during xylitol production by engineered *Escherichia coli*. *Biotechnol. Bioeng.* **102**, 209–220 (2009).

85. Nyyssölä, A., Pihlajaniemi, A., Palva, A., von Weymarn, N., & Leisola, M. Production of xylitol from D-xylose by recombinant *Lactococcus lactis*. *J. Biotechnol.* **118**, 55–66 (2005).

86. Woodyer, R., Simurdiak, M., van der Donk, W.A., & Zhao, H. Heterologous expression, purification, and characterization of a highly active xylose reductase from *Neurospora crassa*. *Appl. Environ. Microbiol.* **71**, 1642–1647 (2005).

87. Nidetzky, B., Helmer, H., Klimacek, M., Lunzer, R., & Mayer, G. Characterization of recombinant xylitol dehydrogenase from *Galactocandida mastotermitis* expressed in *Escherichia coli*. *Chem. Biol. Interact.* **143–144**, 533–542 (2003).

88. Kratzer, R., Pukl, M., Egger, S., & Nidetzky, B. Whole-cell bioreduction of aromatic alpha-keto esters using *Candida tenuis* xylose reductase and *Candida boidinii* formate dehydrogenase co-expressed in *Escherichia coli*. *Microb. Cell Fact.* **7**, 371–412 (2008).

89. Xiong, M., Chen, G., & Barford, J. Alteration of xylose reductase coenzyme preference to improve ethanol production by *Saccharomyces cerevisiae* from high xylose concentrations. *Bioresour. Technol.* **102**, 9206–9215 (2011).

90. Trausinger, G. *et al.* Identification of novel metabolic interactions controlling carbon flux from xylose to ethanol in natural and recombinant yeasts. *Biotechnol. Biofuels* **8**, 157–170 (2015).

91. Watanabe, S. *et al.* Ethanol production from xylose by recombinant Saccharomyces cerevisiae expressing protein engineered NADP$^+$-dependent xylitol dehydrogenase. *J. Biotechnol.* **130**, 316–319 (2007).

92. Bengtsson, O., Hahn-Hagerdal, B., & Gorwa-Grauslund, M.F. Xylose reductase from *Pichia stipitis* with altered coenzyme preference improves ethanolic xylose fermentation by recombinant *Saccharomyces cerevisiae*. *Biotechnol. Biofuels* **2**, 1–10 (2009).

93. Jin, Y.-S., Alper, H., Yang, Y.-T., & Stephanopoulos, G. Improvement of xylose uptake and ethanol production in recombinant *Saccharomyces cerevisiae* through an inverse metabolic engineering approach. *Appl. Environ. Microbiol.* **71**, 8249–8256 (2005).

94. Jin, Y.-S., Jones, S., Shi, N.-Q., & Jeffries, T.W. Molecular cloning of XYL3 (D-Xylulokinase) from *Pichia stipitis* and characterization of its physiological function. *Appl. Environ. Microbiol.* **68**, 1232–1239 (2002).

95. Yu, L., Xu, M., Tang, I.-C., & Yang, S.-T. Metabolic engineering of *Clostridium tyrobutyricum* for n-butanol production through co-utilization of glucose and xylose. *Biotechnol. Bioeng.* **112**, 2134–2141 (2015).

96. Meijnen, J.P., de Winde, J.H., & Ruijssenaars, H.J. Establishment of Oxidative D-Xylose Metabolism in *Pseudomonas putida S12*. *Appl. Environ. Microbiol.* **75**, 2784–2791 (2009).

97. Walfridsson, M., Hallborn, J., Penttilä, M., Keränen, S., & Hahn-Hägerdal, B. Xylose-metabolizing *Saccharomyces cerevisiae* strains overexpressing the Tkl1 and Tal1 genes encoding the pentose-phosphate pathway enzymes transketolase and transaldolase. *Appl. Environ. Microbiol.* **61**, 4184–4190 (1995).

98. Nair, N.U., & Zhao, H. Selective reduction of xylose to xylitol from a mixture of hemicellulosic sugars. *Metab. Eng.* **12**, 462–468 (2010).

99. Wang, X. *et al.* Engineering furfural tolerance in *Escherichia coli* improves the fermentation of lignocellulosic sugars into renewable chemicals. *Proc. Natl. Acad. Sci. USA* **110**, 4021–4026 (2013).

100. Gutierrez, T., Buszko, M.L., Ingram, L.O., & Preston, J.F. Reduction of furfural to furfuryl alcohol by ethanologenic strains of bacteria and its effect on ethanol production from xylose. *Appl. Biochem. Biotechnol.* **98–100**, 327–340 (2002).

101. Pereira, B. *et al.* Efficient utilization of pentoses for bioproduction of the renewable two-carbon compounds ethylene glycol and glycolate. *Metab. Eng.* **34**, 80–87 (2016).

102. Radek, A. *et al.* Engineering of *Corynebacterium glutamicum* for minimized carbon loss during utilization of D-xylose containing substrates. *J. Biotechnol.* **192**, 156–160 (2014).

103. Liu, H., Valdehuesa, K.N.G., Nisola, G.M., Ramos, K.R.M., & Chung, W.-J. High yield production of D-xylonic acid from D-xylose using engineered *Escherichia coli*. *Bioresour. Technol.* **115**, 244–248 (2012).
104. Nair, N.U., & Zhao, H. Evolution in reverse: Engineering a D-Xylose-specific xylose reductase. *Chembiochem* **9**, 1213–1215 (2008).
105. Jeppsson, M. *et al.* The expression of a *Pichia stipitis* xylose reductase mutant with higher KM for NADPH increases ethanol production from xylose in recombinant Saccharomyces cerevisiae. *Biotechnol. Bioeng.* **93**, 665–673 (2006).
106. Petschacher, B., & Nidetzky, B. Altering the coenzyme preference of xylose reductase to favor utilization of NADH enhances ethanol yield from xylose in a metabolically engineered strain of *Saccharomyces cerevisiae*. *Microb. Cell Fact.* **7**, 1–12 (2008).
107. Sriprapundh, D., Vieille, C., & Gregory, Z.J. Molecular determinants of xylose isomerase thermal stability and activity: Analysis of thermozymes by site-directed mutagenesis. *Protein Eng. Des. Sel.* **13**, 259–265 (2000).
108. Wei, N., Quarterman, J., Kim, S.R., Cate, J.H.D., & Jin, Y.-S. Enhanced biofuel production through coupled acetic acid and xylose consumption by engineered yeast. *Nat. Commun.* **4**, 1–8 (2013).
109. Bettiga, M., Bengtsson, O., Hahn-Hagerdal, B., & Gorwa-Grauslund, M.F. Arabinose and xylose fermentation by recombinant *Saccharomyces cerevisiae* expressing a fungal pentose utilization pathway. *Microb. Cell Fact.* **8**, 1–12 (2009).
110. Eiteman, M.A., Lee, S.A., Altman, R., & Altman, E. A substrate-selective co-fermentation strategy with *Escherichia coli* produces lactate by simultaneously consuming xylose and glucose. *Biotechnol. Bioeng.* **102**, 822–827 (2009).
111. Wahlbom, C.F., Cordero Otero, R.R., van Zyl, W.H., Hahn-Hägerdal, B., & Jonsson, L.J. Molecular analysis of a *Saccharomyces cerevisiae* mutant with improved ability to utilize xylose shows enhanced expression of proteins involved in transport, initial xylose metabolism, and the pentose phosphate pathway. *Appl. Environ. Microbiol.* **69**, 740–746 (2003).
112. Yim, H. *et al.* Metabolic engineering of *Escherichia coli* for direct production of 1,4-butanediol. *Nat. Chem. Biol.* **7**, 445–452 (2011).
113. Kim, Y., Ingram, L.O., & Shanmugam, K.T. Construction of an *Escherichia coli* K-12 Mutant for homoethanologenic fermentation of glucose or xylose without foreign genes. *Appl. Environ. Microbiol.* **73**, 1766–1771 (2007).

114. Cordova, L.T., & Antoniewicz, M.R. 13C metabolic flux analysis of the extremely thermophilic, fast growing, xylose-utilizing *Geobacillus* strain LC300. *Metab. Eng.* **33**, 148–157 (2016).

115. Parreiras, L.S. *et al.* Engineering and two-stage evolution of a lignocellu-losic hydrolysate-tolerant *Saccharomyces cerevisiae* strain for anaerobic fermentation of xylose from AFEX pretreated corn stover. *PLoS One* **9**, 1–17 (2014).

Chapter 9

Engineering Metabolism for the Synthesis of Polyhydroxyalkanoate Biopolymers

Guo-Qiang Chen[*,†,‡,§,¶,‖] *and Xiao-Ran Jiang*[*,†,‡]

School of Life Sciences, Tsinghua University, Beijing 100084, China
†*Tsinghua-Peking Center for Life Sciences*
Tsinghua University, Beijing 100084, China
‡*Center for Synthetic and Systems Biology*
Tsinghua University, Beijing 100084, China
§*Center for Nano and Micro Mechanics, Tsinghua University*
Beijing 100084, China
¶*MOE Key Lab of Industrial Biocatalysis, Tsinghua University*
Beijing 100081, China
‖*chengq@mail.tsinghua.edu.cn*

9.1. Introduction

Microbial polyhydroxyalkanoates (PHA) are a family of biodegradable and biocompatible polyesters microbially produced as a source of chemicals, materials, and biofuels.[1] Compared with other well-known biodegradable or bio-based polymers with less CO_2 emission such as polybutylene succinate (PBS) and polylactide (PLA), PHA have much wider diversity in monomers with over 150 structural variations reported,[2] resulting in diverse material properties. These are related to the 14 pathways known (Figure 9.1 and Table 9.1).[3] Therefore, efforts have been made to engineer these pathways, allowing PHA be produced in sufficient amounts with the required structures.

Over the past many years, metabolic engineering researchers have been working hard to improve carbon flux flow toward PHA synthesis, to synthesize diverse PHA, to achieve maximal PHA accumulation, to change the morphology of PHA-accumulating bacteria, and to improve robustness of microorganisms.[4–6]

As shown in Figure 9.1, PHA production is related to many pathways even though pathway I is the most well studied and most representative one. For example, significant changes relating to poly-3-hydroxybutyrate (PHB) production, the most well-studied PHA, were observed in the TCA cycle, lipid synthesis, and amino acid biosynthetic pathways of *Halomonas* sp. KM-1 to shift dramatically between the exponential growth and stationary phases.[7] During the stationary phase, 17 metabolites were observed to be upregulated and a cell dry mass of 18 g/L containing 45% PHB was realized at 24 h in 5% glucose-supplemented cultures, whereas 11 metabolites were upregulated and a cell dry mass of 38 g/L containing 74% PHB was observed at 36 h in 10% glucose-supplemented cultures. This study showed that PHB synthesis is related to many metabolites, and metabolite regulations during PHB accumulation are important for cell growth, indicating that multicomponent and phase-specific mechanisms are involved.

It is therefore important to engineer the above PHA synthesis pathways including glycolysis, TCA cycle, amino acid synthesis, fatty acid synthesis, and β-oxidation, so that desirable PHA with the right structures and properties as well as right amount could be obtained.

9.2. Engineering the Glycolysis Pathway

It was reported that PHB productivity in *Ralstonia eutropha* could be accelerated by extracellular electron transfer (EET) using a biocompatible mediator in an electrochemical system.[8] When the electrode potential was poised at 0.6 V (vs. the standard hydrogen electrode) where the EET pathway was constructed, PHB production rate was enhanced by 60%

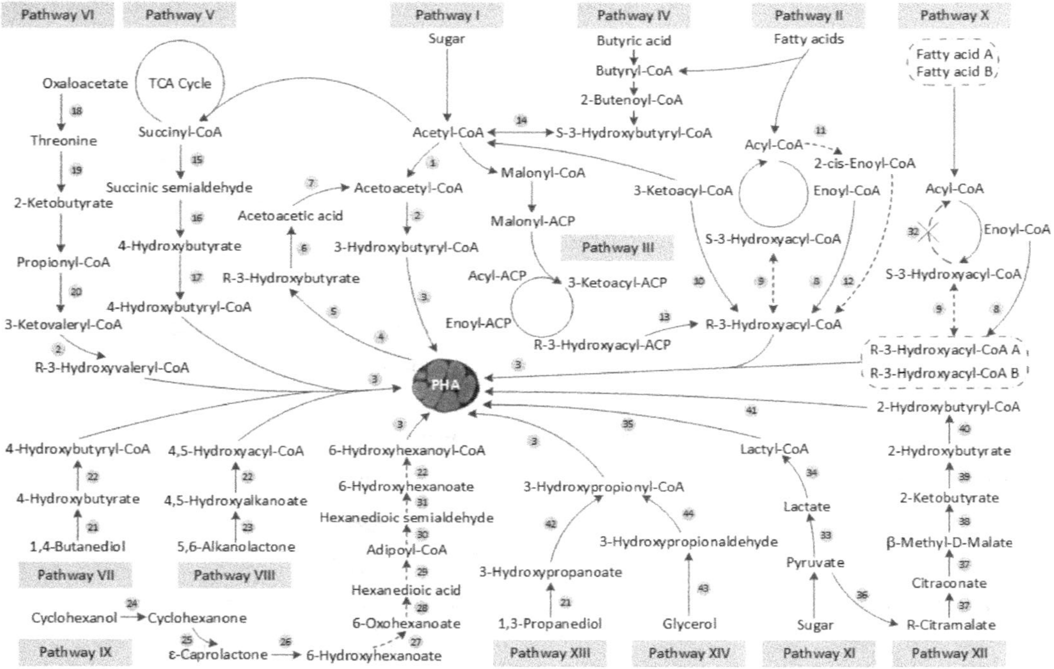

Figure 9.1. Fourteen known PHA pathways.[3] Natural PHA synthesis pathways include: pathway I starts from sugar to form acetyl-CoA, acetoacetyl-CoA to 3-hydroxybutyryl-CoA, which enters the polymerization process to form PHB; Pathway II begins from fatty acid(s) as substrate to enter the β-oxidation cycle, leading to the formation of R-3-hydroxyacyl-CoA monomers for mostly medium-chain-length (mcl) PHA synthesis, while pathway III directs acetyl-CoA to malonyl-CoA to 3-ketoacyl-ACP for forming R-3-hydroxyacyl-CoA monomers. Pathway IV uses butyric acid without entering the β-oxidation cycle to form S-3-hydroxybutyryl-CoA and then acetyl-CoA. The types of PHA formed depend not only on monomer supply pathways, but also on the specificity of PHA synthases. Generally, a low specificity of a PhaC allows the formation of diverse PHA structures. There are engineering pathways leading to unconventional PHA. These are pathways V to XIV.[3]

Table 9.1. PHA production related enzymes.[3]

No.	Pathway	Abbreviation	Enzyme	Species
1	Pathway I	PhaA	β-Ketothiolase	*R. eutropha*
2		PhaB	NADPH-dependent acetoacetyl-CoA reductase	
3		PhaC	PHA synthase	
4	Associated way	PhaZ	PHA depolymerase	*Aeromonas hydrophila* 4AK4
5			Dimer hydrolase	*Pseudomonas stutzeri* 1317
6			(R)-3-Hydroxybutyrate dehydrogenase	*R. eutropha*
7			Acetoacetyl-CoA synthase	*Pseudomonas oleovorans*
8	Pathway II	PhaJ	(R)-Enoyl-CoA hydratase/enoyl-CoA hydratase I	*Pseudomonas aeruginosa*
9			Epimerase	*A. hydrophila* 4AK4
10		FabG	3-Ketoacyl-CoA reductase	*Pseudomonas putida* KT2442
11			Acyl-CoA oxidase, putative	
12			Enoyl-CoA hydratase, putative	
13	Pathway III	PhaG	3-Hydroxyacyl-ACP-CoA transferase	*Pseudomonas mendocina*
		FadD	Malonyl-CoA-ACP transacylase	recombinant *Escherichia coli*
14	Pathway IV		NADH-dependent acetoacetyl-CoA reductase	*Rhizobium* (Ciser) sp. CC 1192
15	Pathway V	SucD	Succinic semialdehyde dehydrogenase	*Clostridium kluyveri*

16		4hbD	4-Hydroxybutyrate dehydrogenase	
17		OrfZ	4-Hydroxybutyrate-CoA transferase	
18	Pathway VI	ThrA	Aspartokinase I	E. coli
		ThrB	Homoserine kinase	
		ThrAC	Threonine synthase	
19		IlvA	Threonine deaminase	
20		BktB(PhaA)		
21	Pathway VII	DhaT	Alcohol dehydrogenase	P. putida KT2442
		AldD	Aldehyde dehydrogenase	
22			Hydroxyacyl-CoA synthase, putative	Mutants and recombinant of Alcaligenes eutrophus
23	Pathway VIII		Lactonase, putative	
24	Pathway IX	ChnA	Cyclohexanol dehydrogenase	Acinetobacter sp. SE19
25		ChnB	Cyclohexanone monooxygenase	Brevibacterium epidermidis HCU
26		ChnC	Caprolactone hydrolase	
27		ChnD	6-Hydroxyhexanoate dehydrogenase	
28		ChnE	6-Oxohexanoate dehydrogenase	
29			Semialdehyde dehydrogenase, putative	

(Continued)

Table 9.1. *(Continued)*

No.	Pathway	Abbreviation	Enzyme	Species
30			6-Hydroxyhexanoate dehydrogenase, putative	
31			Hydroxyacyl-CoA synthase, putative	
32	Pathway X	FadA (deleted)	3-Ketoacyl-CoA thiolase	*Pseudomonas entomophila*
		FadB (deleted)	3-hydroxyacyl-CoA dehydrogenase	
33	Pathway XI	ldhA	Lactate dehydrogenase	*E. coli*
34		Pct$_{Cp}$	Propionate CoA-transferase	*Clostridium propionicum*
35		PhaC1$_{Ps6-19}$	Type II PHA synthase	*Pseudomonas* sp. MBEL 6–9
36	Pathway XII	CimA	α-Isopropylmalate synthase	*Leptospira interrogans*
37		LeuCD	3-Isopropylmalate dehydratase	
38		LeuB	3-Isopropylmalate dehydrogenase	
39		PanE	2-Hydroxybutyrate dehydrogenase	*L. lactis* subsp. *lactis* Il1403
40		Pct540$_{Cp}$	Propionate CoA-transferase	recombinant *E. coli*
41		PhaC1437$_{Ps6-19}$	Type II PHA synthase	
42	Pathway XIII	Pcs'	Propionyl-CoA synthase	*Chloroflexus aurantiacus*
43	Pathway XIV	DhaB	Glycerol dehydratases	*Klebsiella pneumoniae*
44		PduP	Propionaldehyde dehydrogenase	*Salmonella typhimurium*

compared to the case without the EET process. This result points to the fact that the extracellular anode served as an additional electron acceptor, leading to the acceleration of glycolysis and hence PHB synthesis,[8] indicating that engineering glycolysis improves PHB synthesis.

Glycolysis has been generally accepted to supply the reducing power for the anaerobic conversion of volatile fatty acids (VFAs) to PHA by polyphosphate-accumulating organisms (PAOs). However, the importance of the TCA cycle has also been raised since 1980s. To achieve this goal, the glycogen pool of an activated sludge highly enriched in *Candidatus accumulibacter,* a putative PAO was reduced substantially through starving the sludge under intermittent anaerobic and aerobic conditions. Acetate added was still taken up anaerobically and stored as PHA under this condition, with negligible glycogen degradation.[9]

To increase the reducing power supply for enhanced PHB synthesis, NAD kinase was overexpressed in recombinant *E. coli* harboring PHB synthesis pathway via an accelerated supply of NADPH, which is one of the most crucial factors influencing PHB production. Shake flask studies revealed that excess NAD kinase in *E. coli* harboring the PHB synthesis operon increased the accumulation of PHB to 16–35% compared with the controls alone with an increase of NADP concentration up to 6-fold. Under the same growth conditions without process optimization, the NAD kinase-overexpressing recombinant produced 14 g/L PHB compared with 7 g/L produced by the control in a 28-h fermentor study.[10]

Substrate to PHB yield, Y(PHB/glucose), increased from 0.08 g PHB/g glucose for the control to 0.15 g PHB/g glucose for the NAD kinase-overexpressing strain, a 76% increase for the Y(PHB/glucose). These results demonstrated that the overexpression of NAD kinase can be used for improving PHB synthesis.[10]

9.3. Engineering the TCA Cycle

Proteomic and transcriptomic studies showed that active PHA production coordinated with the TCA cycle to maintain balanced growth in wild-type *Haloarcula hispanica* grown in nutrient-limited medium (PHA-accumulating conditions) vs. nutrient-rich medium (non-PHA-accumulating conditions).[11] Under nutrient-limiting conditions with an excess carbon source, PHA synthetic genes including *phaEC, phaB*, and *phaP* were upregulated at the transcriptional level, whereas the TCA cycle and respiratory chain were downregulated. Thus, acetyl-CoA could be fed into the PHA pathway, leading to PHA accumulation.[11]

A large amount of NADPH required during PHA synthesis was likely supplied by the C3 (pyruvate) and C4 (malate) pathway coupled with the urea cycle. When PHA synthesis was blocked, namely, in the PHA synthase mutant vs. wild type grown in nutrient-limited medium, the mutant could direct additional carbon and energy to the TCA cycle without obvious contribution to biomass accumulation.[11]

A recombinant *E. coli* was constructed for poly(3-hydroxybutyrate-*co*-4-hydroxybutyrate) [P(3HB-*co*-4HB)] synthesis from glucose.[12] Genes involved in succinate degradation in *Clostridium kluyveri* and PHB accumulation pathway of *R. eutropha* were coexpressed. At the same time, *E. coli* native succinate semialdehyde dehydrogenase genes *sad* and *gabD* were both deleted for eliminating succinate formation from succinate semialdehyde, enhancing the carbon flux to 4HB biosynthesis. The resulting *E. coli* produced 9 g/L cell dry weight containing 66% P(3HB-*co*-11 mol% 4HB) using glucose as the carbon source in a 48 h shake flask growth. In a fermentor study, a 24 g/L cell dry weight containing 63% P(3HB-*co*-12.5 mol% 4HB) was obtained after 29 h of cultivation.[12]

Clustered regularly interspaced short palindromic repeats interference (CRISPRi) was used to control the PHA biosynthesis pathway flux and to adjust PHA composition. A pathway was constructed in *E. coli* for the production of P(3HB-*co*-4HB) from glucose. The native gene *sad* encoding *E. coli* succinate semialdehyde dehydrogenase was expressed under the control of CRISPRi using five specially designed single-guide RNAs (sgRNAs) for regulating carbon flux to 4-hydroxybutyrate (4HB) synthesis. P(3HB-*co*-4HB) consisting of 1–9 mol% 4HB was produced. Additionally, succinate, generated by succinyl-CoA synthetase and succinate dehydrogenase (respectively, encoded by genes *sucC, sucD* and *sclhAl, sdhB*), was channeled preferentially to the 4HB precursor by using selected sgRNAs such as *sucC2, sucD2, sclhB2*, and *sclhAl* via CRISPRi (Figure 9.2), forming P(3HB-*co*-14 to 18 mol% 4HB) depending on the expression levels of the downregulated genes. CRISPRi is a

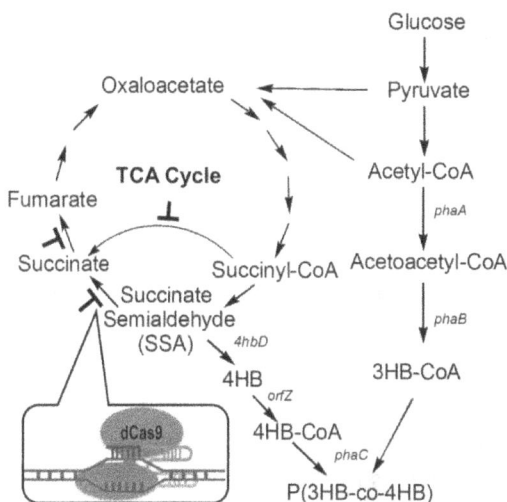

Figure 9.2. Engineering TCA cycle for P(3HB-*co*-4HB) synthesis via CRISPRi.[13] CRISPRi is a tool to control the P3HB4HB biosynthesis pathway flux and to adjust 3HB/4HB composition formed by recombinant *E. coli*. The CRISPRi system was used to repress gene transcription initiation and elongation in the related pathways in TCA cycle. To obtain P3HB4HB consisting of various 4HB ratios, several genes can be manipulated simultaneously, including the following genes: *phaA*, *β*-ketothiolase; *phaB*, NADPH-dependent acetoacetyl-CoA reductase; *phaC*, PHA synthase; *sucD*, succinate semi-aldehyde dehydrogenase; *4hbD*, 4-hydroxybutyrate dehydrogenase; *orfZ*, CoA transferase, and so on (Table 9.1).

feasible method to simultaneously manipulate multiple genes in TCA cycle of *E. coli*.[13]

9.4. Engineering the Amino Acid Synthesis Pathway

Corynebacterium crenatum SYPA 5 is an industrial strain for L-arginine production. The introduction of the PHB synthesis pathway into several strains can regulate the global metabolic pathway. The microbial pathways of PHB and L-arginine biosynthesis are NADPH-dependent. NAD kinase could upregulate the NADPH concentration in the bacteria. *Corynebacterium crenatum* P1-containing PHB synthesis pathway was constructed and cultivated in batch fermentation for 96 h. Key enzyme activities were observed to increase compared to the control strain *C. crenatum* SYPA 5. More PHB was found in *C. crenatum* P1, up to 12.7% of the dry cell weight. Higher growth level and enhanced glucose consumptions were also observed in *C. crenatum* P1. The yield of L-arginine was increased by 21% compared to control under the influence of PHB accumulation.

Accumulation of PHB by introducing PHB synthesis pathway, together with upregulation of coenzyme level by overexpressing NAD kinase, allows the recombinant *C. crenatum* to become a high-efficiency cell factory for L-arginine production.[14]

Lin *et al.* reported the discovery of a new pathway called threonine bypass by flux balance analysis of the genome-scale metabolic model of *E. coli*.[15] It mainly contains reactions for threonine synthesis and degradation. The pathway can potentially increase PHB yield and other acetyl-CoA-derived products by reutilizing the CO_2 released at the pyruvate dehydrogenase step. The threonine and serine degradation pathways were deregulated with enhanced threonine synthesis, resulting in more than 2-fold improvement of the PHB titer. When *glyA* gene was overexpressed for enhancing glycine to serine conversion combined with activated transhydrogenase, the resulting strain produced 7 g/L PHB with a yield of 0.36 g/g glucose in the shake flask studies and 36 g/L PHB with a yield of 0.23 g/g glucose in a fed-batch fermentation, which was more than 3-fold higher than the parent strain.

9.5. Engineering the β-Oxidation Pathway

Pseudomonas putida KT2442 produces medium-chain-length PHAs consisting of 3-hydroxyhexanoate (3HHx), 3-hydroxyoctanoate (3HO), 3-hydroxydecanoate (3HD), 3-hydroxydodecanoate (3HDD), and 3-hydroxytetradecanoate (3HTD) from relevant fatty acids or from *in situ* fatty acid synthesis pathway (Figure 9.3). *P. puitda* KT2442 was found to contain key fatty acid degradation enzymes encoded by genes *PP2136, PP2137 (fadB* and *fadA)*, and *PP2214, PP2215 (fadB2x* and *fadAx)*, respectively (Figure 9.4).[16]

The above-mentioned enzymes and other important fatty acid degradation enzymes including 3-hydroxyacyl-CoA dehydrogenase and

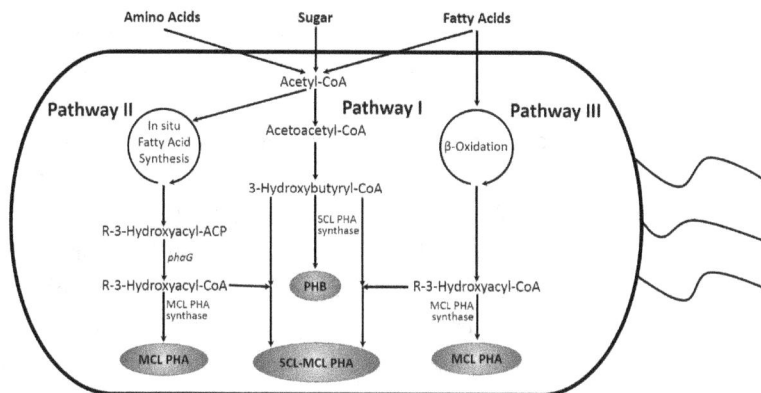

Figure 9.3. The three major PHA synthesis mechanisms (pathways).[4] Pathway I, The Acetyl-CoA to 3-hydroxybutyryl-CoA pathway; Pathway II, The *in situ* fatty acid synthesis pathway; Pathway III, The β-oxidation pathway.

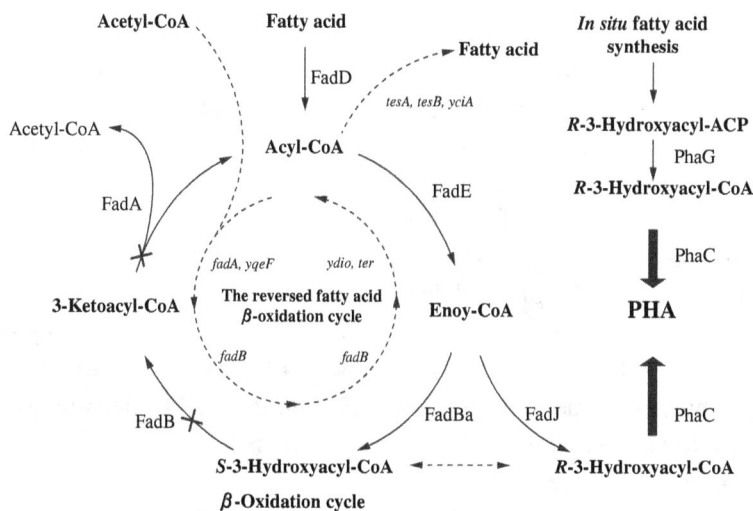

Figure 9.4. Engineering the β-oxidation pathway for diversifying PHA structures.[4] Deleting partial β-oxidation cycle to control PHA compositions. Enzymes in β-oxidation cycle: FadD, fatty acid-CoA ligase; FadE, acyl-CoA dehydrogenase; FadBa, *S*-enoyl-CoA hydratase; FadB, 3-hydroxyacyl-CoA dehydrogenase; FadA, acetyl-CoA acetyltransferase; PhaJ, R-enoyl-CoA hydratase; PhaC, PHA synthase; PhaG, 3-hydroxyacyl-CoA-acyl carrier protein transferase; Genes in the reversed fatty acid β-oxidation cycle, *yqeF/fadA*, thiolase; *fadB*, hydroxyacyl-CoA dehydrogenase/enoyl-CoA hydratase; *ydiO*, enoyl-CoA reductase; *ter*, trans-2-enoyl-CoA reductase; *tesA/tesB/yciA*, thioesterase.

acyl-CoA dehydrogenase encoded by genes *PP2047* and *PP2048*, respectively, were deleted to investigate how their affected PHA structures. Mutant *P. puitda* KTQQ20 was constructed containing the above six deleted genes and also 3-hydroxyacyl-CoA-acyl carrier protein transferase encoded by PhaG (Figure 9.4). A significant reduction on fatty acid β-oxidation activity was observed with the mutant. *P. puitda* KTQQ20 synthesized homopolymer poly-3-hydroxydecanoate (PHD) or P(3HD-*co*-84 mol% 3HDD) when grown on decanoic acid or dodecanoic acid, demonstrating that homopolymer PHD and 3HDD monomers dominating PHA were successfully synthesized by β-oxidation inhibiting *P. putida* grown on relevant carbon sources.[16]

A mutant was constructed from *P. putida* KT2442 by deleting its *phaG* gene encoding *R*-3-hydroxyacyl-ACP-CoA transacylase and several other β-oxidation related genes including *fadB, fadA, fadB2x,* and *fadAx*. The resulting *P. putida* KTHH03 was able to synthesize homopolymers including poly(3-hydroxyhexanoate) (PHHx) and poly(3-hydroxyheptanoate) (PHHp), together with a near homopolymer poly(3-hydroxyoctanoate-*co*-2 mol% 3-hydroxyhexanoate) (PHO*) in presence of hexanoate, heptanoate, and octanoate, respectively. When the PHA synthase genes *phaC1* and *phaC2* were deleted, the resulting *P. putida* KTHH08 containing PHA synthesis operon *phaPCJ* from *Aeromonas hydrophila* 4AK4 accumulated homopolymer poly(3-hydroxyvalerate) (PHV) when valerate was used as a carbon source. The *phaC* deleted *P. putida* KTHH06-harboring PHA synthase, PhbC, from *Ralstonia eutropha*-produced homopolymers, poly(3-hydroxybutyrate) (PHB) and poly(4-hydroxybutyrate) (P4HB), using γ-butyrolactone as a precursor. *P. putida* KT2442 derivatives is a platform to produce various PHA homopolymers.[17]

PHA synthesis genes *phaPCJ* cloned from *Aeromonas caviae* were transformed into *P. putida* KTOY06, a mutant of *P. putida* KT2442, to produce a short-chain-length and medium-chain-length PHA block copolymer consisting of PHB as one block and random copolymer of 3-hydroxyvalerate (3HV) and 3-hydroxyheptanoate (3HHp) as another block. Compared with other commercially available PHA, the short- and medium-chain length block copolymer PHB-*b*-PHVHHp enjoyed improved mechanical properties.[18] Another diblock copolymers consisting of PHB block covalently bonded with poly-3-hydroxyhexanoate (PHHx) block were for the first time produced successfully by a recombinant *P. putida* KT2442 with its β-oxidation cycle deleted to its maximum. In comparison to a random copolymer poly-3-hydroxybutyrate-*co*-3-hydroxyhexanoate [P(HB-*co*-HHx)] and a blend sample of PHB and PHHx, the PHB-*b*-PHHx showed improved structural-related mechanical properties.[19]

A medium-chain-length (mcl) PHA producer *Pseudomonas entomophila* L48 was found to produce mcl PHA consisting of 3-hydroxyhexanoate (3HHx), 3-hydroxyoctanoate (3HO), 3-hydroxydecanoate (3HD), and 3-hydroxydodecanoate (3HDD) from related carbon-source fatty acids. Some of the genes encoding key enzymes in β-oxidation cycle of *P. entomophila* such as 3-hydroxyacyl-CoA dehydrogenase, 3-ketoacyl-CoA thiolase, and acetyl-CoA acetyltransferase were deleted to construct mutant *P. entomophila* LAC26, which accumulated over 90% PHA consisting of 99 mol% 3HDD. The new type of PHA also represented high crystallinity. For the first time, P(3HDD) homopolymers were obtained.[20]

Microbial synthesis of functional polymers has become increasingly important for industrial biotechnology. It is now possible to synthesize controllable composition of poly(3-hydroxyalkanoate) (PHA) consisting of 3-hydroxydodecanoate (3HDD) and phenyl group on the side-chain when β-oxidation of *P. entomophila* was weakened. In the presence of 5-phenylvaleric acid (PhV), the mutated *P. entomophila* synthesized homopolymer poly(3-hydroxy-5-phenylvalerate) or P(3HPhV). While copolyesters P(3HPhV-*co*-3HDD) of 3-hydroxy-5-phenylvalerate (3HPhV) and 3-hydroxydodecanoate (3HDD) were synthesized when grown on mixtures of phenylvaleric acid and dodecanoic acid, compositions of 3HPhV in P(3HPhV-*co*-3HDD) ranged from 3% to 32% depending on dodecanoic acid/5-phenylvaleric acid ratios. The results demonstrated the possibility of tailor-made novel functional PHA using β-oxidation-weakened *P. entomophila*.[21]

Functional PHAs allow chemical modifications to widen the PHA diversity, promising to increase the value of these biodegradable and biocompatible polyesters. Unsaturated PHA side chains can be easily grafted to add chemical groups, and to cross-link with other PHA polymer chains. The β-oxidation weakened *P. entomophila*[21] synthesized random copolymers of 3-hydroxydodecanoate (3HDD) and 3-hydroxy-9-decenoate (3H9D). P(3HDD-*co*-3H9D) structures can be adjusted by ratios of dodecanoic acid (DDA) to 9-decenol (9DEO) fed to the culture of *P. entomophila*. Homopolymer P3H9D was formed when only 9DEO was added to the culture. Diblock copolymers of P3HDD-*b*-P3H9D was produced by feeding DDA as the first precursor to form P3HDD block followed by adding 9DEO as the second precursor to form the second P3H9D block.[22]

The random copolymers P(3HDD-*co*-3H9D) could be cross-linked under UV radiation due to the presence of the unsaturated bonds. It was found that the diblock polymer P3HDD-*b*-P3H9D increased at least 2-fold on Young's modulus compared with its random copolymers consisting of similar 3HDD/3H9D ratios. This study demonstrated that PHA functionality could be controlled to meet various requirements.[22]

9.6. Conclusion

Metabolic engineering approaches allow a lot of possibilities regarding PHA synthesis, such as formation of homopolymers, random copolymers, block copolymers, and functional polymers. In addition, we can increase PHA synthesis by channeling more carbon flux toward PHA synthesis, or deleting β-oxidation, which reduced our fatty acid substrate utilization efficiency. In addition, the expression of PHA synthesis pathway can improve the production of some amino acids. This increases the value of PHA synthesis. Over time, PHA applications can be developed to meet various requirements.

References

1. Gao, X., Chen, J.C., Wu, Q., & Chen, G.Q. Polyhydroxyalkanoates as a source of chemicals, polymers, and biofuels. *Curr. Opin. Biotechnol.* **22**, 768–774 (2011).
2. Steinbüchel, A., & Valentin, H.E. Diversity of bacterial polyhydroxyalkanoic acid. *FEMS Microbiol. Lett.* **128**, 219–228 (1995).

3. Meng, D.C. *et al.* Engineering the diversity of polyesters. *Curr. Opin. Biotechnol.* **29**, 24–33 (2014).

4. Chen, G.Q., Hajnal, I., Wu, H., Lv, L., & Ye, J. Engineering biosynthesis mechanisms for diversifying polyhydroxyalkanoates. *Trends Biotechnol.* **33**, 565–574 (2015).

5. Jiang, X.R., & Chen, G.Q. Morphology engineering of bacteria for bioproduction. *Biotechnol. Adv.* **34**, 435–440 (2016).

6. Chen, G.Q., & Hajnal, I. The 'PHAome'. *Trends Biotechnol.* **33**, 559–564 (2015).

7. Jin, Y.X., Shi, L.H., & Kawata, Y. Metabolomics-based component profiling of *Halomonas sp.* KM-1 during different growth phases in poly(3-hydroxybutyrate) production. *Bioresour. Technol.* **140**, 73–79 (2013).

8. Nishio, K. *et al.* Extracellular electron transfer enhances polyhydroxybutyrate productivity in *Ralstonia eutropha*. *Environ. Sci. Technol. Lett.* **1**, 40–43 (2014).

9. Zhou, Y., Pijuan, M., Zeng, R.J., & Yuan, Z. Involvement of the TCA cycle in the anaerobic metabolism of polyphosphate accumulating organisms (PAOs). *Water Res.* **43**, 1330–1340 (2009).

10. Li, Z.J., Cai, L., Wu, Q., & Chen, G.Q. Overexpression of NAD kinase in recombinant *Escherichia coli* harboring the *phbCAB* operon improves poly(3-hydroxybutyrate) production. *Appl. Microbiol. Biotechnol.* **83**, 939–947 (2009).

11. Liu, H. *et al.* Proteome reference map of *Haloarcula hispanica* and comparative proteomic and transcriptomic analysis of polyhydroxyalkanoate biosynthesis under genetic and environmental perturbations. *J. Proteome Res.* **12**, 1300–1315 (2013).

12. Li, Z.J. *et al.* Production of poly(3-hydroxybutyrate-co-4-hydroxybutyrate) from unrelated carbon sources by metabolically engineered *Escherichia coli*. *Metab. Eng.* **12**, 352–359 (2010).

13. Lv, L., Ren, Y.L., Chen, J.C., Wu, Q., & Chen, G.Q. Application of CRISPRi for prokaryotic metabolic engineering involving multiple genes, a case study: Controllable P(3HB-co-4HB) biosynthesis. *Metab. Eng.* **29**, 160–168 (2015).

14. Xu, M. *et al.* Effect of Polyhydroxybutyrate (PHB) storage on L-arginine production in recombinant *Corynebacterium crenatum* using coenzyme regulation. *Microb. Cell Fact.* **15**, 15 (2016).

15. Lin, Z. *et al.* Metabolic engineering of *Escherichia coli* for poly(3-hydroxybutyrate) production via threonine bypass. *Microb. Cell Fact.* **14**, 185 (2015).

16. Liu, Q., Luo, G., Zhou, X.R., & Chen, G.Q. Biosynthesis of poly(3-hydroxydecanoate) and 3-hydroxydodecanoate dominating polyhydroxyalkanoates by β-oxidation pathway inhibited *Pseudomonas putida. Metab. Eng.* **13**, 11–17 (2011).

17. Wang, H.H., Zhou, X.R., Liu, Q., & Chen, G.Q. Biosynthesis of polyhydroxyalkanoate homopolymers by *Pseudomonas putida. Appl. Microbiol. Biotechnol.* **89**, 1497–1507 (2011).

18. Li, S.Y., Dong, C.L., Wang, S.Y., Ye, H.M., & Chen, G.Q. Microbial production of polyhydroxyalkanoate block copolymer by recombinant *Pseudomonas putida. Appl. Microbiol. Biotechnol.* **90**, 659–669 (2011).

19. Tripathi, L., Wu, L.P., Chen, J., & Chen, G.Q. Synthesis of Diblock copolymer poly-3-hydroxybutyrate-block-poly-3-hydroxyhexanoate [PHB-b-PHHx] by a β-oxidation weakened *Pseudomonas putida* KT2442. *Microb. Cell Fact.* **11**, 44 (2012).

20. Chung, A.L. *et al.* Biosynthesis and characterization of poly(3-hydroxydodecanoate) by β-oxidation inhibited mutant of *Pseudomonas entomophila* L48. *Biomacromolecules* **12**, 3559–3566 (2011).

21. Shen, R. *et al.* Benzene containing polyhydroxyalkanoates homo- and copolymers synthesized by genome edited *Pseudomonas entomophila. Sci. China. Life Sci.* **57**, 4–10 (2014).

22. Li, S. *et al.* Microbial synthesis of functional homo-, random, and block polyhydroxyalkanoates by β-oxidation deleted *Pseudomonas entomophila. Biomacromolecules* **15**, 2310–2319 (2014).

Index

www.ingramcontent.com/pod-product-compliance
Lightning Source LLC
Chambersburg PA
CBHW050550190326
41458CB00007B/1993

* 9 7 8 1 7 8 6 3 4 4 2 9 8 *